Theorems And Problems In Functional Analysis - The Answer Book I: Elements of Set Theory and Topology

Detailed solutions of the exercises in Kirillov's and Gvichiani's Theorems and Problems in Functional Analysis

Martin Rupp

January 1, 2013

~~~~~~~~~~~~~~~~

USE OF THE ANSWER BOOK

~~~~~~~~~~~~~~~

For students, self-learners and any audience who wish to self-train and improve their ability in mathematics and problem solving, we strongly suggest that you start by reading the core book, "Theorem and Problems in Functional Analysis" (TPFA) ([?] and [?]) by Kirillov and Gvichiani, then select exercises that fit your needs and/or your interest. If you are having difficulty on a specific problem and find the hint provided by TPFA confuses you even more, then you should start to read the solution of the corresponding exercises. If you have solved the exercise and you want to compare your result with the solution, it is also a very good idea to read the corresponding section. Conversely it would be unwise to read this book without first trying to solve the problems.

For Teachers, and other educationalists who wish to find useful core material in order to submit problems to their students , these volumes might be an interesting tool, and can be used as "raw" bricks to build original problems or can be used as an "old-fashioned" ready-for-use self-contained "Answer Book" in conjunction with TPFA.

DIFFICULTY CODE FOR THE EXERCISES:

$\mathscr{B}$: in general fairly easy to solve but often requires rigorous attention so as not to be trapped by its apparent simplicity.

$-$: in general solvable but might require good understanding and practice in the theory of TPFA and some background knowledge.

¶: fairly complicated, will need good general mathematics background combined with very good understanding of the theory in TPFA and will require insight to be solved.

¶¶: hard to solve, will need insight and a good intuition (or luck, searches in books treating the subject,...), expert problem solving techniques outside the range of the theory provided by TPFA.

Preface

About the 5 Volumes.

The goal of the 5 Volumes of **Theorems and Problems in Functional Analysis - The Answer Book** is to provide the complete solutions for the exercises compiled in the "core book" by A.A Kirillov and A.Gvichiani, **"Theorems and Problems in Functional Analysis**, MIR editions, 1979" (**TPFA**). The problems cover most of the concepts of classical Functional Analysis: *Distributions, Hilbert Spaces, Harmonic Analysis, Spectral Theory...* The *828* exercises provided in **TPFA** represent an incredibly rich source of problems in order to train oneself in these fields. These exercises also bring numerous original entry points for various subjects outside the field of the theory developed in the core book. Finally, it is important to note that while the volumes are obviously naturally bundled with the core book, they have been written completely independently of it.

- We respect as far as possible the original notations and their conventions.

- We always expand the solutions as much as possible.

- We have deliberately chosen not to make systematic references to **TPFA** and *not to reproduce the hints* (whilst sometimes we provide extra-hints).

- We almost always follow the direction of the hints found in **TPFA** but we also provide additional or alternative proofs and interesting extended references to other topics linked with the exercises.

About this volume.

Volume I deals with the first $23 \times 3 = 69$ exercises of the core book and covers *Set Theory, Orders Relationships and Category Theory*. The reader will find several intriguing problems about the Möbius functions and their use in Number Theory, the construction of Free Groups and Tensor Algebras from a Category Theory point of view, and p-adics fields properties.

Chernivtsi,   *Martin Rupp*
June 2011

Website.

`http://tpfa-answer-book.info`

Correspondence.

Exercise	Russian Version	English Version
33	33	-
34	34	33
34bis	-	34

All other exercises have identical numeration.

Notations.

$\#A, |A|$: number of elements of the finite set A.

$\mathscr{L}(E, F)$: space of linear mappings from a vector space E to a vector space F.

$span(x_i)$: vector space spanned by the family x_i.

$< x_i >$ may refer to the vector space, algebra or group spanned by the family x_i depending of the context.

$(p, q) \leq n$ means that $p \leq n$ and $q \leq n$ and $(x_n, y_n) \to a$ means that both sequences $\{x_n\}$ and $\{y_n\}$ tends towards a when $n \to \infty$.

$[|a, b|]$, for $(a, b) \in \mathbb{Z}^2$ is the set $\{c \in \mathbb{Z}, a \leq c \leq b\}$.

$p^{\mathbb{N}}$ or p^{∞} is the set of sequences $\{x_n\}_{n \geq 0}$ such that $x_n \in [|0, p-1|]$ for $p \in \mathbb{N}$.

$[x]$, $x \in \mathbb{R}$ usually refers to the integer part of x.

$\sim$ may refer to an equivalence relationship, an homeomorphism or an homomorphism depending of the context.

$Mor(A, B)$ refers to the morphism between two objects A and B of a category.

$GL_n(\mathbb{R})$ refers to the general linear group.

$A \vee B$ means A or B, $A \wedge B$ means A and B.

χ_A refers usually to the characteristic function of A.

π may refer to the set of prime numbers.

$[a, b)$ represents the set of points $\{x, a \leq x < b\}$.

A set A which is enumerable or countable means usually that there exists an injection from A into $\mathbb{N}$. When A is such that $\#A < \infty$ this is usually mentioned or considered to be implicit from the context.

$Hom(A, B)$ represents the set of homomorphisms from a group A into a group B.

$*$ may refer to the convolution operation.

Programs.

Perl programs may be run with any free distribution of PERL version 5 available at:

```
http://www.perl.org.
```

To Pharoah and Philemo.n
To all the errants whoever they are, dogs
or humans, Jews or Gipsies, Nomads or Soldiers.

Acknowledgements

I am indebted to Thomas Langlet for his critical reviewing of some of the solutions. I would like also to express my gratitude to Andre and Doreen Winter from Nottingham for having reviewed the English language.

Contents

Exercises 47 to 69
Categories and Functors   175

Exercises 1 to 23
Relationships - The choice axiom and Zorn's Lemma

- I met Max Zorn once. I was taking graduate level Real Analysis as part of my PhD studies at Indiana University. Of course, I had just learned Zorn's lemma. I assumed that it was made up by someone long dead.

It was a day with dark foreboding clouds. Just as I walked into the snack machine room, there was a sudden massive downpour. I saw an elderly gentleman struggling to close the window. I went over and helped. He thanked me. We shook our heads together at the suddenness of the storm.

After he walked out, a classmate said, "Dude, that was Max Zorn, of Zorn's Lemma."

Jimbo Wales

1/828 $\mathscr{B}$

Subject: Equivalence Relationships

Among the following relationships, indicate those who are equivalence relationships:

1. equality of two numbers;

2. similitude of two triangles;

3. order relationship on the real set;

4. linear dependency in a vector space L of dimension $n \geq 1$;

5. linear dependency in the space $L^* = L \setminus \{0\}$, where L is a vector space.

SOLUTION:

1) The equality $=$ of two numbers is an equivalence relationship because for numbers x, y, z:

$$x = x,$$

$$x = y \implies y = x,$$

$$x = y = z \implies x = z.$$

2) The similitude $\sim$ of two triangles is an equivalence relationship because for triangles $ABC, A'B'C', A''B''C''$:

$$ABC \sim ABC,$$

$$ABC \sim A'B'C' \implies A'B'C' \sim ABC,$$

$$ABC \sim A'B'C' \sim A''B''C'' \implies ABC \sim A''B''C''.$$

3) The order relation on $\mathbb{R}$, is not an equivalence relationship. As a matter of fact, this relation is reflexive and transitive because, for $(x,y,z) \in \mathbb{R}$, we have:

$$x \geq x,$$

$$x \geq y \geq z \implies x \geq z.$$

But it is not a symmetric relationship because:

$$x \geq y \wedge y \geq x \implies x = y.$$

4) In a n-vector space E, let say that $x \sim y$ if x is linearly dependent with y. If $x \in E$, then $x - x = 0$ so x is linearly dependent with x and we have reflexivity:

$$x \sim x$$

if $x \sim y$ so that $\alpha x + \beta y = 0, \alpha \vee \beta \neq 0$ then $\beta y + \alpha x = 0, \beta \vee \alpha \neq 0$ and so we have symmetry:

$$x \sim y \implies y \sim x$$

if $x \sim y \sim z$ then $\alpha x + \beta y = 0, \alpha \vee \beta \neq 0$ and $\alpha' y + \beta' z = 0, \alpha' \vee \beta' \neq 0$ so that $\alpha \alpha' x - \beta \beta' z = 0$ but we can have both $\alpha \alpha' = 0$ and $\beta \beta' = 0$ (for example $\alpha = 0$ and $\beta' = 0$) so that we do not have necessarily $x \sim z$ [1], so transitivity is not always true and thus the linear dependency in a n-vector space is not an equivalence relationship.

5) The previous results remain valid so that we have reflexivity and symmetry. If $\alpha \alpha' x - \beta \beta' z = 0$, then $\alpha \alpha' \neq 0$ or $\beta \beta' \neq 0$ so that we have transitivity and so, linear dependency in L^* is an equivalence relationship.

---

[1]This means concretely that we have always $x \sim 0$ and $y \sim 0$ for any two $(x,y) \in E \times E$ but not necessarily $x \sim y$

2/828

Subject: Limits and Equivalence Relationships

We will say that two strictly positive functions f_1, f_2 on $[0, 1]$ are equivalent if

$$0 < \underline{lim}_{x \to 0} \frac{f_1(x)}{f_2(x)}, \overline{lim}_{x \to 0} \frac{f_1(x)}{f_2(x)} < \infty.$$

Check that it is an equivalence relationship and that the corresponding quotient set is non-enumerable.

SOLUTION:

We will name this relationship $\sim$. We assume that $f_1 \sim f_2$ if $\exists (a, b, \varepsilon) \in (\mathbb{R}^+ - 0) \times (\mathbb{R}^+ - 0) \times (\mathbb{R}^+ - 0)$ such that $a < \frac{f_1(x)}{f_2(x)} < b, \forall x \in [0, \varepsilon[$.

As a matter of fact, if we have this property, then $0 < a \le \liminf_{x \to 0} \frac{f_1}{f_2}$ and $\limsup_{x \to 0} \frac{f_1}{f_2} \le b < \infty$ so that $f_1 \sim f_2$.

Conversely, if $f_1 \sim f_2$, we put $a = \liminf_{x \to 0} \frac{f_1}{f_2}$ and $b = \limsup_{x \to 0} \frac{f_1}{f_2}$, from the definition of upper and lower limit, we can find ε_1 such that $a \le \frac{f_1(x)}{f_2(x)}, \forall x \in [0, \varepsilon_1[$ and ε_2 such that $\frac{f_1(x)}{f_2(x)} \le b, \forall x \in [0, \varepsilon_2[$. We put $\varepsilon = \inf(\varepsilon_1, \varepsilon_2)$ and since we can find a', b' such that $0 < a' < a$ and $b < b' < \infty$, the property will be verified.

It is obvious that $f_1 \sim f_1$ (reflexivity).

If $f_1 \sim f_2$ then $a < \frac{f_1(x)}{f_2(x)} < b, \forall x \in [0, \varepsilon[$ we have $a^{-1} < \frac{f_2(x)}{f_1(x)} < b^{-1}, \forall x \in [0, \varepsilon[, a^{-1} > 0, b^{-1} > 0$ and therefore $f_2 \sim f_1$ (symmetry).

If $f_1 \sim f_2 \sim f_3$ then we have:

$$a < \frac{f_1(x)}{f_2(x)} < b, \forall x \in [0, \varepsilon[, a > 0, b > 0$$

and:

$$a' < \frac{f_2(x)}{f_3(x)} < b', \forall x \in [0, \varepsilon'[, a' > 0, b' > 0$$

so that:

$$aa' < \frac{f_1(x)}{f_3(x)} < bb', \forall x \in [0, inf(\varepsilon, \varepsilon')[, aa' > 0, bb' > 0$$

and thus $f_1 \sim f_3$ (transitivity).

We have checked that $\sim$ is an equivalence relationship over the set of functions from $[0,1]$ onto $(\mathbb{R}^+ - 0)$. Let us call $\mathbf{Q}$ the corresponding quotient set. In order to prove the innumerability, we note that there is an injection F from $\mathbb{R}^+ - 0$ into $\mathbf{Q}$ defined by:

$$F(\alpha) = \overline{x^\alpha}$$

where $\overline{x^\alpha}$ denotes the class over $\mathbf{Q}$ of the positive functions $x \to x^\alpha$.

If $\alpha > \beta$, $F(\alpha) \neq F(\beta)$ because $\liminf_{x \to 0} \frac{x^\alpha}{x^\beta} = \limsup_{x \to 0} \frac{x^\alpha}{x^\beta} = \lim_{x \to 0} x^{\alpha - \beta} = 0$ so that F is an injection.

3/828

Subject: A partial order for positive functions

We define the relation $f_1 \succ f_2$ for strictly positive functions on $[0, 1]$ if:

$$f_1 \succ f_2 \iff \lim_{x \to 0} \frac{f_1(x)}{f_2(x)} = \infty.$$

Check that this is a partial order relationship and that all enumerable subsets are bounded.

SOLUTION:

This relation is anti-symmetric because it is not possible to have both $f_1 \succ f_2$ and $f_2 \succ f_1$ and so $R \cap R' = \emptyset \subset \Delta_x{}^2$. If $f_1 \succ f_2 \succ f_3$ then:

$$\lim_{x \to 0} \frac{f_1(x)}{f_2(x)} = \infty,$$

$$\lim_{x \to 0} \frac{f_2(x)}{f_3(x)} = \infty$$

and, by multiplying the above two equations:

$$\lim_{x \to 0} \frac{f_1(x)}{f_3(x)} = \infty \times \infty = \infty.$$

So that $f_1 \succ f_3$ (transitivity).

---

[2]See [?] for notations

We consider a set of strictly positive functions $(f_i)_{i \in A}$, where A is a subset of $\mathbb{N}$, and we define the following positive functions:

$$\underline{f}(x) = [x^{-1}]^{-1} \min_{1 \leq i \leq [x^{-1}]} f_i(x), \overline{f}(x) = [x^{-1}] \max_{1 \leq i \leq [x^{-1}]} f_i(x)$$

where $[x]$ is the integer part of x.
We consider a $j \in A$:

$$\frac{\overline{f}(x)}{f_j(x)} = [x^{-1}] \frac{\max_{1 \leq i \leq [x^{-1}]} f_i(x)}{f_j(x)}.$$

We can find ε_j such that $x < \varepsilon_j \implies j < [x^{-1}]$. Then, for such values of x:

$$\frac{\overline{f}(x)}{f_j(x)} \geq [x^{-1}].$$

So that we can conclude:

$$\lim_{x \to 0} \frac{\overline{f}(x)}{f_j(x)} = \infty.$$

Or, equivalently:

$$\overline{f} \succ f_i, \forall i \in A.$$

For $x < \varepsilon_j$, we have also:

$$\frac{\underline{f}(x)}{f_j(x)} = \frac{1}{[x^{-1}]} \frac{\min_{1 \leq i \leq [x^{-1}]} f_i(x)}{f_j(x)},$$

$$\frac{\underline{f}(x)}{f_j(x)} \leq \frac{1}{[x^{-1}]}.$$

So that we can conclude:

$$\lim_{x \to 0} \frac{\underline{f}(x)}{f_j(x)} = 0.$$

Or, equivalently:

$$\underline{f} \prec f_i, \forall i \in A.$$

So that:

$$\overline{f} \succ f_i \succ \underline{f}, \forall i \in A.$$

4/828 $\mathscr{B}$

Subject: A partial order relationship on a product of set

We consider X and Y, two partially ordered sets. We define a relationship on $X \times Y$ by:

$$(x_1, y_1) \geq (x_2, y_2) \Leftrightarrow x_1 \geq x_2 \wedge y_1 \geq y_2.$$

Prove that it is a partial order relationship.
If X and Y are ordered sets, is it an order?

SOLUTION:

We have $\{(x_1, y_1) \geq (x_2, y_2)\} \wedge \{(x_2, y_2) \geq (x_1, y_1)\} \implies \{x_1 \geq x_2 \wedge y_1 \geq y_2\} \wedge \{x_2 \geq x_1 \wedge y_2 \geq y_1\} \Leftrightarrow \{(x_1 \geq x_2 \wedge x_2 \geq x_1)\} \wedge \{(y_2 \geq y_1 \wedge y_1 \geq y_2)\} \implies x_1 = x_2 \wedge y_1 = y_2.$

So we have:

$$\{(x_1, y_1) \geq (x_2, y_2)\} \wedge \{(x_2, y_2) \geq (x_1, y_1)\} \implies (x_1, y_1) = (x_2, y_2).$$

This means that the relation R (e.g. the relation $\geq$) is such that: $R \cap R' \subset \Delta_x$ (antisymmetry).

If we have $(x_1, y_1) \geq (x_2, y_2) \geq (x_3, y_3)$, then:

$$x_1 \geq x_2 \geq x_3, y_1 \geq y_2 \geq y_3.$$

So that:

$$x_1 \geq x_3, y_1 \geq y_3.$$

And, finally, we have transitivity:

$$(x_1, y_1) \geq (x_3, y_3).$$

This shows that the relationship is a partial order.

If we suppose that X and Y are (totally) ordered, then the relation R we defined is not necessarily a (total) order relation. We can use the example of $\mathbb{N}^2$: $\mathbb{N}$ is a totally ordered set but if we take, for example $(1,4)$ and $(4,1)$, these two elements cannot be compared in $\mathbb{N}^2$ by using our order product R.

5/828

Subject: Product of partially ordered sets

a) Let $(X_\alpha)_{\alpha \in A}$ a family of partially ordered sets. We provide their direct product $\prod_{\alpha \in A} X_\alpha$ with the relation $\succeq$ defined by: $(x = \prod_{\alpha \in A} x_\alpha) \succeq (y = \prod_{\alpha \in A} y_\alpha)$ if $x_\alpha \geq y_\alpha, \forall \alpha \in A$.

Show that $\succeq$ is a partial order. The product $\prod_{\alpha \in A} X_\alpha$ provided with this order is named *product of partially ordered sets*.

b) We distinguish a point $x_\alpha \in X_\alpha$ and we call product of the couples $(X_\alpha; x_\alpha)$ the subset $\prod_{\alpha \in A}(X_\alpha; x_\alpha) \subset \prod_{\alpha \in A} X_\alpha$ made of collections (y_α) so that the set $\{\alpha, x_\alpha \neq y_\alpha\}$ is finite. Provide $\prod_{\alpha \in A}(X_\alpha; x_\alpha)$ with a partial order.

c) Prove that the set of natural numbers partially ordered by divisibility | with the point 1 distinguished is isomorph to the product of an enumerable product of $\mathbb{N}$ provided with the natural order and with 0 as a distinguished point.

SOLUTION:

a) Let us show antisymmetry: if $x \succeq y$ and $y \succeq x$ for x and y in the product set, then, if $x = \prod_{\alpha \in A} x_\alpha$ and $y = \prod_{\alpha \in A} y_\alpha$ we have:

$$x_\alpha \geq y_\alpha, \forall \alpha \in A, y_\alpha \geq x_\alpha, \forall \alpha \in A.$$

As every set inside the product is partially ordered, this means that:

$$x_\alpha = y_\alpha, \forall \alpha \in A.$$

That is to say: $x = y$.

If we have: $x \succeq y \succeq z$ then, with the same notations:

$$x_\alpha \geq y_\alpha \geq z_\alpha, \forall \alpha \in A.$$

Again, as every set inside the product is partially ordered, this means that:

$$x_\alpha \geq z_\alpha, \forall \alpha \in A.$$

That is to say $x \succeq z$ (transitivity).

This shows that $\succeq$ is a partially ordered set.

b) Now, we provide $\prod_{\alpha \in A}(X_\alpha; x_\alpha)$ with a partial order:

$(y_\alpha, x_\alpha)_{\alpha \in A} \succeq (y'_\alpha, x_\alpha)_{\alpha \in A}$ if $(y_\alpha)_{\alpha \in A} \succeq (y'_\alpha)_{\alpha \in A}$ as follows:

we consider the relation $\succeq$ induced on $\prod_{\alpha \in A}(X_\alpha; x_\alpha)$ by the partial order defined in a) over $\prod_{\alpha \in A} X_\alpha$.

If we consider $X = \prod_{\alpha \in A} X_\alpha$ and $x = \prod_{\alpha \in A} x_\alpha$ the distinguished point of X then for two elements $y = \prod_{\alpha \in A} y_\alpha$ and $z = \prod_{\alpha \in A} z_\alpha$ in $(X; x)$ there are only a finite number of y_α's and z_α's that are different to x_α. Let us call B this set of α's:

$$y \preceq z \Leftrightarrow \forall \alpha \in B, z_\alpha \leq y_\alpha.$$

c) We can consider, respectively, $(\mathbb{N}, \times)$ and $(\mathbb{N}, +)$ as, respectively, a multiplicative and an additive monoid. Thus $\prod_{i=1}^{\infty}(\mathbb{N}, +)$ will be also an additive monoid (provided with the monoid product structure). If we consider $\prod_{i=1}^{\infty}((\mathbb{N}, +); 0) = (\prod_{i=1}^{i=\infty}(\mathbb{N}, +); \prod_{i=1}^{i=\infty}\{0\}))$, then, again this will be an additive monoid.

If we define the map φ from the multiplicative monoid with a distinguished point $((\mathbb{N}, \times); 1)$ onto $(\prod_{i=1}^{i=\infty}(\mathbb{N}, +); \prod_{i=1}^{i=\infty}\{0\}))$ by: $\varphi(n) = (m_1, \ldots, m_k, 0, \ldots, 0, \ldots)$ where $m_k = max\{i \in \mathbb{N}, (p_k)^i | n, p_k \in \pi$ (p_k being the kth prime number and π being the set of prime numbers) then we see that φ is a monoid homomorphism.

Indeed one check that $\varphi(1) = (0, \ldots, 0, \ldots)$ so that the distinguished point from $(\mathbb{N}, \times)$ is mapped to the distinguished point from $\prod_{i=1}^{\infty}(\mathbb{N}, +)$.

If $n = \prod_{p_k \in \pi} p_k^{m_k}$ and $n' = \prod_{p_k \in \pi} p_k^{m'_k}$, where n and n' are any two number in $\mathbb{N}$ then:

$\varphi(nn') = \{(m_k + m'_k)\}_{k=1}^{k=\infty} = \{(m_k)\}_{k=1}^{k=\infty} + \{(m'_k)\}_{k=1}^{k=\infty} = \varphi(n) + \varphi(n')$.

From the unicity of decomposition by prime numbers in $\mathbb{N}$, we can see that φ is bijective so that φ is in fact a monoid-isomorphism.

Now if we partially order $(\mathbb{N}, \times)$ by divisibility and if we order $(\mathbb{N}, +)$ by the natural order, we see that:

$$n | n' \Leftrightarrow (\forall k > 0), m_k \leq m'_k \Leftrightarrow \varphi(n) \leq \varphi(n').$$

So that φ preserve the order.

The set $(\prod_{i=1}^{i=\infty}(\mathbb{N}, +); \prod_{i=1}^{i=\infty}\{0\})$ is isomorphic as a monoid with a distinguished point partially ordered with the product order from the natural order to $((\mathbb{N}, \times); 1)$ as a monoid with a distinguished point partially ordered by divisibility.

6/828 $\mathscr{B}$

Subject: Definition of a Cauchy sequence by means of generalized sequences

Provide a definition of Cauchy sequences using generalized[3] sequences $d_{m,n} = d(x_m, x_n)$.

SOLUTION:

We consider $\mathbb{N}^2 = \mathbb{N} \times \mathbb{N}$ provided with the order product $\preceq$ obtained from the product of the natural order of $\mathbb{N}$ (see Exercise 4). Let us suppose that (X, d) is a metric space and that $\{x_n\}_{n \geq 0}$ is a sequence over X. First we note that $\mathbb{N}^2$ provided with the order product is a filtering set. Indeed, if $(a, b) \in \mathbb{N}^2$ and $(c, d) \in \mathbb{N}^2$ then we can always find (α, β) such that[4]:

$$(a, b) \preceq (\alpha, \beta),$$
$$(c, d) \preceq (\alpha, \beta).$$

We can then define the generalized sequence $\{d_\alpha\}_{\alpha \in \mathbb{N}^2}$ over (X, d) by:

$$d_\alpha = d(x_m, x_n), \alpha = (m, n).$$

---

[3] See [?] for the definition of a generalized sequence and of a filtering set
[4] We only need to make $\alpha \geq \max(a, c), \beta \geq \max(b, d)$

Now, we claim that $\{x_n\}_{n\geq 0}$ is a Cauchy sequence if and only if $\alpha \longrightarrow \infty \Rightarrow d_\alpha \longrightarrow 0$ [5].

Indeed, $(m,n) \to \infty \Rightarrow d(x_m, x_n) \to 0$ which is equivalent to: $\alpha \longrightarrow \infty \Rightarrow d_\alpha \longrightarrow 0$.

---

[5]We use the symbol $\longrightarrow$ to indicate that this is the convergence of a generalized sequence

7/828 $\mathcal{B}$

Subject: Partial order on the set of parts of a set

Let us define $\mathscr{D}(X)$ as the power set of X partially ordered by inclusion. If $X = \bigsqcup_{\alpha \in A} X_\alpha$ then show that $\mathscr{D}(X)$ is isomorph to $\prod_{\alpha \in A} \mathscr{D}(X_\alpha)$.

SOLUTION:

Let $\mathscr{D}(X)$ be the power set of X ordered by the inclusion $\subset$.

If $X = \bigsqcup_{\alpha \in A} X_\alpha$, then $\mathscr{D}(X) \approx \prod_{\alpha \in A} \mathscr{D}(X_\alpha)$ ($\approx$ meaning here being equivalent as partially ordered sets, e.g. we can find a strictly increasing bijection between the two sets).

If we define, for $Y \in \mathscr{D}(X)$, the set Y_α by $Y_\alpha = Y \cap X_\alpha$, then we have $Y = \bigsqcup_{\alpha \in A} Y_\alpha$. Y is being uniquely defined by its "coordinates" Y_α over each X_α, e.g. $Y = (Y_\alpha)_{\alpha \in A}$.

Furthermore, if $Y = (Y_\alpha)_{\alpha \in A}$ and $Z = (Z_\alpha)_{\alpha \in A}$, then $Y \subset Z \Leftrightarrow Y_\alpha \subset Z_\alpha$ foreach $\alpha \in A$.

The set of these coordinates is $\prod_{\alpha \in A} \mathscr{D}(X_\alpha)$ so that we may conclude to the equivalence between the two sets.

Alternatively, we may also note that, for a set X, the power set of X is equivalent to 2^X, the space of functions $X \to \{0, 1\}$. Indeed, this equivalence is the association $Y \to \chi_Y$, where χ_Y is the characteristic function of Y.

Next we also note that, then $\prod_{\alpha \in A} 2^{X_\alpha}$ is also equivalent to $2^{\bigsqcup_{\alpha \in A} X_\alpha}$, $\bigsqcup_{\alpha \in A} X_\alpha$ being the disjunctive union $\bigvee_{\alpha \in A} X_\alpha$.

Indeed if $\{f_\alpha\}_{\alpha \in A}$ is a set of functions defined by: $f_\alpha : X_\alpha \to \{0,1\}$, then this set is also equivalent to a single function over $\bigsqcup_{\alpha \in A} X_\alpha$ given by $f = \sum_{\alpha \in A} f_\alpha$.

Then we may also conclude since $2^X \approx 2^{\bigsqcup_{\alpha \in A} X_\alpha} \approx \prod_{\alpha \in A} 2^{X_\alpha} \approx \prod_{\alpha \in A} \mathscr{D}(X_\alpha)$.

8/828

Subject: The Möbius function

We consider $(X, \geq)$ a partially ordered set. We suppose that the partial order of X has the following property: the set $M(x) = \{y \in X, y \leq x\}$ is finite for all $x \in X$. For every function $f(x)$ over X we define:

$$F(x) = \sum_{y \leq x} f(y).$$

Show that $f(x)$ can be computed from $F(x)$ by the following formula:

$$f(x) = \sum_{y \leq x} \mu(x, y) F(y)$$

where $\mu(x, y)$ is a function independent of f, uniquely defined by the ordered set X.

μ is called **the Möbius function of** $(X, \geq)$.

SOLUTION:

FUNDAMENTAL PROPERTY OF MÖBIUS FUNCTIONS.

Let us calculate:

$$S_\mu(x, y) = \sum_{y \leq z \leq x} \mu(z, y).$$

From:

$$F(x) = \sum_{z \leq x} f(z)$$

and:

$$f(z) = \sum_{y \leq z} \mu(z,y) F(y),$$

we deduce that:

$$F(x) = \sum_{y \leq z \leq x} \mu(z,y) F(y),$$

$$F(x) = \sum_{y, y \leq z \leq x} F(y) \left(\sum_{z, y \leq z \leq x} \mu(z,y) \right) = \sum_{y, y \leq x} F(y) S_\mu(x,y).$$

If the previous equation is to be true for all functions f, we must have:

$$S_\mu(x,y) = \delta_y(x) \ ^6.$$

We know that $M(x) = \{y \in X, y \leq x\}$ is a finite set, let us put $M(x) = \{x_i\}_{i=1,\ldots,N}$, we can reorder the x_i so that $x_j \geq x_i \implies j \geq i$ [7].

We note that, if for example $M(x) = M(x_N) = \{x_1, x_2, \ldots, x_N\}$, we have: $M(x_1) \subseteq M(x_N) \subseteq \ldots M(x_{N-1}) \subseteq M(x_1)$.

From:

$$F(x) = \sum_{y \leq x} f(y).$$

We deduce:

---

[6] δ is the **krönecker function**: $\delta_y(x) = 0$ if $y \neq x$, $\delta_y(y) = 1$)
[7] This will be proved in Exercise 22

$$F(x_N) = \sum_{y \in M(x_N)} f(y) = f(x_1) + \ldots + f(x_N),$$

$$F(x_{N-1}) = \sum_{y \in M(x_{N-1})} f(y) = \varepsilon_{1,N-1} f(x_1) + \ldots + f(x_{N-1}),$$

$$\ldots$$

$$F(x_i) = \sum_{y \in M(x_i)} f(y) = \varepsilon_{1,i} f(x_1) + \ldots + \varepsilon_{k,i} f(x_k) + \ldots + f(x_i),$$

$$\ldots$$

$$F(x_1) = \sum_{y \in M(x_1)} f(y) = f(x_1).$$

Where $\varepsilon_{k,i}$ is being defined by $\varepsilon_{k,i} = \chi_{M(x_k)}(x_i) \in \{0; 1\}$.

It has to be noticed that we cannot necessarily compare any x_i with an x_j but anyway $M(x_i)$ cannot have elements x_j with $j > i$.

Now, the N-vector:

$$\tilde{F} = \begin{pmatrix} F(x_1) \\ \ldots \\ F(x_N) \end{pmatrix}$$

and the N-vector:

$$\tilde{f} = \begin{pmatrix} f(x_1) \\ \ldots \\ f(x_N) \end{pmatrix}$$

are linked by the following equation:

$$\tilde{F} = \mathbf{Z}\tilde{f}$$

where $\mathbf{Z}$ is the N-square matrix defined by:

$$\mathbf{Z} = \begin{pmatrix} 1 & \ldots & 1 & 1 & 1 \\ \varepsilon_{1,N-1} & \ldots & \varepsilon_{N-2,N-1} & 1 & 0 \\ \varepsilon_{2,N-1} & \ldots & 1 & 0 & 0 \\ \ldots & \ldots & \ldots & \ldots & \\ 1 & 0 & 0 & 0 & 0 \end{pmatrix}$$

$\mathbf{Z}$ is a triangular matrix with a determinant $\det(\mathbf{Z}) = 1$ so that $\mathbf{Z}$ is invertible:

$$\widetilde{f} = \mathbf{Z}^{-1}\widetilde{F}$$

and thus $\mathbf{Z}^{-1}$ defines the function μ by the following formula:

$$\mathbf{Z}^{-1} =$$

$$\begin{pmatrix} \mu(x_N,x_1) & \cdots & \mu(x_N,x_{N-2}) & \mu(x_N,x_{N-1}) & \mu(x_N,x_N) \\ \chi_{M(x_{N-1})}(x_1) \times \mu(x_{N-1},x_1) & \cdots & \chi_{M(x_{N-1})}(x_{N-2}) \times \mu(x_{N-1},x_1) & \mu(x_{N-1},x_{N-1}) & 0 \\ \chi_{M(x_{N-2})}(x_1) \times \mu(x_{N-2},x_1) & \cdots & \mu(x_{N-2},x_{N-2}) & 0 & 0 \\ \cdots & \cdots & \cdots & \cdots & \cdots \\ \mu(x_1,x_1) & \cdots & 0 & 0 & 0 \end{pmatrix}$$

which we read:

$$f(x_N) =$$
$$\sum_{y \in M(x_N)} \mu(x,y)F(y) = \mu(x_N,x_1)F(x_1) + \ldots + \mu(x_N,x_N)F(x_N),$$

$$f(x_{N-1}) =$$
$$\sum_{y \in M(x_{N-1})} \mu(x,y)F(y) = \chi_{M(x_{N-1})}(x_1) \times \mu(x_{N-1},x_1)F(x_1) + \ldots$$
$$+\mu(x_{N-1},x_{N-1})F(x_{N-1}),$$

$$\ldots$$

$$f(x_1) =$$
$$\sum_{y \in M(x_{N-1})} \mu(x,y)F(y) = \mu(x_1,x_1)F(x_1).$$

And this is exactly the definition of the Möbius function.

9/828

Subject: Möbius function of a product $\prod_\alpha (X_\alpha; x_\alpha)$

We consider a family $(X_\alpha; x_\alpha)$ of partially ordered sets X_α with distinguished points x_α. We suppose that every X_α satisfies the condition of Exercise 8 - that is to say that the Möbius function μ_α may be defined over X_α. We suppose that, for all α, except maybe a finite number of them, x_α is the minimum element of X_α.

Show that on the partially ordered set $\prod_{\alpha \in A}(X_\alpha; x_\alpha)$, we can define the Möbius function μ by the formula:

$$\mu(\{y_\alpha\}, \{z_\alpha\}) = \prod_{\alpha \in A} \mu_\alpha(y_\alpha, z_\alpha).$$

SOLUTION:

From exercise 5) we have defined the product of sets with a distinguished point $\prod_{\alpha \in A}\{X_\alpha; x_\alpha\}$ as the family of $\{y_\alpha\}_{\alpha \in A}$ such that $y_\alpha \neq x_\alpha$ only for a finite number of α's.

We can define the function μ over $\prod_{\alpha \in A}\{X_\alpha; x_\alpha\}$ by:

$$\mu(\{y_\alpha\}_{\alpha \in A}, \{z_\alpha\}_{\alpha \in A}) = \prod_{\alpha \in A} \mu_\alpha(y_\alpha, z_\alpha),$$

(where μ_α is the Möbius function of X_α).

Indeed, since $y_\alpha \neq x_\alpha$ and $z_\alpha \neq x_\alpha$ only for a finite number of α's, let us say for all α's in B, we have:

$$\prod_{\alpha \in A} \mu_\alpha(y_\alpha, z_\alpha) = \prod_{\alpha \in B} \mu_\alpha(y_\alpha, z_\alpha) \times \prod_{\alpha \in A-B} \mu_\alpha(x_\alpha, x_\alpha),$$

$$= \prod_{\alpha \in B} \mu_\alpha(y_\alpha, z_\alpha),$$

$$= \prod_{i=1}^{i=k} \mu_\alpha(y_{\alpha_i}, z_{\alpha_i}).$$

So that the product is finite and thus well-defined.
From here we note that:

$$S = \sum_{y \leq z \leq u} \prod_{\alpha \in A} \mu_\alpha(y_\alpha, z_\alpha),$$

$$= \sum_{\{y_\alpha\}_{\alpha \in A} \leq \{z_\alpha\}_{\alpha \in A} \leq \{u_\alpha\}_{\alpha \in A}} \prod_{\alpha \in A} \mu_\alpha(y_\alpha, z_\alpha),$$

$$= \sum_{\{y_\alpha\}_{\alpha \in A} \leq \{z_\alpha\}_{\alpha \in A} \leq \{u_\alpha\}_{\alpha \in A}} \prod_{\alpha \in C} \mu_\alpha(y_\alpha, z_\alpha) \times \sum_{\{y_\alpha\}_{\alpha \in A} \leq \{z_\alpha\}_{\alpha \in A} \leq \{u_\alpha\}_{\alpha \in A}} \prod_{\alpha \in A-C} \mu_\alpha(y_\alpha, z_\alpha).$$

Where C is defined as the (finite) set where $y_\alpha \neq x_\alpha$ or $z_\alpha \neq x_\alpha$ or $x_\alpha \neq \min X_\alpha$.

When α is in $C-A$, then the set of $(y_\alpha, z_\alpha, u_\alpha)$'s such that $y_\alpha \leq z_\alpha \leq u_\alpha$ is reduced to a single point, namely $(x_\alpha, x_\alpha, x_\alpha)$ and $\mu_\alpha(y_\alpha, z_\alpha) = \mu_\alpha(x_\alpha, x_\alpha) = 1$ so that we find out that:

$$S = \sum_{\{y_\alpha\}_{\alpha \in A} \leq \{z_\alpha\}_{\alpha \in A} \leq \{u_\alpha\}_{\alpha \in A}} \prod_{\alpha \in C} \mu_\alpha(y_\alpha, z_\alpha).$$

C is a finite set so that the sum S is well-defined and we can factorize the sum of products into a product of sums, e.g.

$$S = \prod_{\alpha \in C} \sum_{y_\alpha \leq z_\alpha \leq u_\alpha} \mu_\alpha(y_\alpha, z_\alpha),$$

$$= \prod_{\alpha \in C} \delta(y_\alpha, z_\alpha),$$

$$= \delta(y, z).$$

Therefore μ verifies the fundamental property of Möbius functions and then μ is a Möbius function over $\prod_{\alpha \in A} \{X_\alpha; x_\alpha\}$.

10/828 ¶¶

Subject: Calculation of various Möbius functions

Find the Möbius function of the following partially ordered sets:
1. The natural series with the usual order;

2. The natural series with the divisibility order;

3. The set of the finite parts of a set ordered by inclusion;

4. The set of the parts of a vector space of dimension n over a finite field $\mathbb{K}$ ordered by inclusion.

SOLUTION:

1) We consider $(\mathbb{N}, <)$. We have[8]:

$$S_\mu(x,y) = \sum_{y \leq z \leq x} \mu(z,y) = \delta_y(x).$$

So that, by taking $y = x$:

$$\mu(x,x) = 1.$$

By taking $y = x - 1$:

$$\mu(x-1,x-1) + \mu(x,x-1) = 0.$$

---

[8]See Exercise 8

So that:

$$\mu(x,x-1) = -1.$$

For $y < x-1$: from $S_\mu(x,y) = 0$ and $S_\mu(x-1,y) = 0$, we get:

$$\mu(y,y) + \mu(y+1,y) + \mu(y+2,y) + \ldots + \mu(x-1,y) + \mu(x,y) = 0.$$

Or:

$$\mu(y,y) + \mu(y+1,y) + \ldots + \mu(x-1,y) + \mu(x,y) = 0,$$
$$\mu(y+1,y) + \ldots + \mu(x-1,y) + \mu(x,y) = 0.$$

And this implies $\mu(x,y) = 0$.

That means that, over $(\mathbb{N}, <)$, the Möbius function is defined by:

$$\begin{aligned}
\mu(x,x) &= 1, \\
\mu(x,x-1) &= -1, \\
\forall y \le x-2, \mu(x,y) &= 0.
\end{aligned}$$

2) Let us now consider $(\mathbb{N}, |)$. If p is a prime, then S_μ is defined by:

$$S_\mu(p,y) = \sum_{y|z|p} \mu(z,y).$$

So that we have only two possibilities: either $y = p$, or $y = 1$:

$$S_\mu(p,p) = \sum_{y|z|p} \mu(z,y) = \mu(p,p) = 1$$

and:

$$S_\mu(p,1) = \sum_{1|z|p} \mu(z,1) = \mu(1,1) + \mu(p,1) = 0.$$

What leads to $\mu(p,1) = -1$.
So that if $p \in \pi$ [9], we have:

$$\mu(p,p) = 1, \mu(p,1) = -1.$$

Now, in the general case, we note also that, if $y|x$, then $\mu(x,y) = \mu(\frac{x}{y},1)$. As we know, μ is uniquely defined by the system:

$$S_\mu(x,y) = \sum_{y|z|x} \mu(z,y) = \delta_y(x)$$

but we have also equivalently:

$$S_{\widetilde{\mu}}(x,y) = \sum_{y|z|x} \widetilde{\mu}(z,y) = \delta_y(x),$$

where $\widetilde{\mu}$ is defined by:

$$\widetilde{\mu}(z,y) = \mu(\frac{z}{y},1).$$

Since the Möbius function is unique, we must have: $\widetilde{\mu} = \mu$.
Let us prove now that μ is a multiplicative function, e.g. that:

$$\Delta(P,Q) = 1 \implies \mu(PQ,1) = \mu(P,1) \times \mu(Q,1).$$

We know that[10]:

$$F(x) = \sum_{y|x} f(y) = (f * 1)(x).$$

$f(x)$ can be computed from $F(x)$ by the following formula:

$$f(PQ) = \sum_{y|PQ} \mu(PQ,y)F(y) = \sum_{y|PQ} \mu(PQ,ab)F(ab) = \sum_{a|P,b|Q} \mu(\frac{PQ}{ab},1)F(ab),$$

---

[9] We note π the set of prime numbers
[10] Where the operation $*$ is the *convolution* in $(\mathbb{N},|)$

$$f(P) = \sum_{a|P} \mu(P,a)F(a)$$

and:

$$f(Q) = \sum_{b|P} \mu(Q,b)F(b).$$

Now, let us suppose that f is an additive function. Since F is defined by $F = f * 1$, F is also additive[11].

$$f(P)f(Q) = \sum_{a|P} \mu(P,a)F(a) \sum_{b|P} \mu(Q,b)F(b) = \sum_{a|P,b|Q} \mu(P,a)\mu(Q,b)F(a)F(b),$$

and:

$$f(P)f(Q) = \sum_{a|P,b|Q} \mu(\frac{P}{a},1)\mu(\frac{P}{b},1)F(a)F(b)$$

for $(a,b) = 1, \forall(a,b)$ such that $a|P$ and $b|Q$ so that $F(a)F(b) = F(ab)$.

We have:

$$f(P)f(Q) = \sum_{a|P,b|Q} \mu(\frac{P}{a},1)\mu(\frac{P}{b},1)F(ab)$$

and:

$$f(PQ) = \sum_{a|P,b|Q} \mu(\frac{PQ}{ab},1)F(ab).$$

We have $f(P)f(Q) = f(PQ)$ and thus: $\mu(\frac{P}{a},1)\mu(\frac{P}{b},1) = \mu(\frac{PQ}{ab},1)$.
μ is additive so that for $x = \prod_{i=1,...,N} p_i^{\alpha_i}$, we have:

---

[11] We will admit that the convolution product of two additive functions is additive, this is a basic number theory result

$$\mu(x,1) = \prod_{i=1,\dots,N} \mu(p_i^{\alpha_i},1).$$

In order to calculate $\mu(p^\alpha)$, p a prime > 1, we use the fact that $S_\mu(P,1) = \delta_1(P), \forall P$:

$$S_\mu(p,1) = \sum_{a|p} \mu(a,1) = 0,$$

$$\mu(1,1) + \mu(p,1) = 0.$$

So that: $\mu(p,1) = -1$

$$S_\mu(p^\alpha,1) = \sum_{a|p^\alpha} \mu(a,1) = 0.$$

$\forall \alpha > 1$, we have:

$$\sum_{i=0,\dots,\alpha} \mu(p^i,1) = \mu(1,1) + \mu(p,1) + \dots + \mu(p^\alpha,1) = \mu(p^2,1) + \dots + \mu(p^\alpha,1) = 0.$$

So that:

$$\mu(p^\alpha,1) = 0, \alpha > 1.$$

So, the values of μ in $(\mathbb{N},|)$ are defined by:
a) $\mu(1) = 1$,
b) for all square-free numbers $x = p_1 \times \dots \times p_k, p_i \in \pi, p_i \neq p_j, \forall (i,j) \in [|1,k|]^2$:

$$\mu(p_1 \times \dots \times p_k, 1) = (-1)^k,$$

c) $\mu(x) = 0$ if x is not square free and $x > 1$.
3) We consider the finite set E, partially ordered by the inclusion. Again, we want to prove that μ has the following properties:

$$\mu(A,B) = \mu(A - B, \emptyset).$$

μ is multiplicative for elements X, Y such that $X \cap Y = \emptyset$:

$$\mu(X \sqcup Y, \emptyset) = \mu(X, \emptyset)\mu(Y, \emptyset).$$

We use the same method as shown in b):
μ is uniquely defined by the system:

$$S_\mu(A, B) = \sum_{B \subseteq Z \subseteq A} \mu(Z, B) = \delta_A(B)$$

but we have also equivalently:

$$S_{\tilde{\mu}}(A, B) = \sum_{B \subseteq Z \subseteq A} \tilde{\mu}(Z, B) = \delta_A(B)$$

where $\tilde{\mu}$ is defined by:

$$\tilde{\mu}(Z, B) = \mu(Z - B, \emptyset).$$

Again, because the Möbius function is unique, we must have: $\tilde{\mu} = \mu$.
In order to prove the multiplicativity of μ, we note that:

$$F(X) = \sum_{Y \subseteq X} f(Y).$$

$f(X)$ can be computed from $F(X)$ by the following formula:

$$
\begin{aligned}
f(G \sqcup H) &= \sum_{Z \subseteq G \sqcup H} \mu(G \sqcup H, Z) F(Z), \\
&= \sum_{K \subseteq G, L \subseteq H} \mu(G \sqcup H, K \sqcup L) F(K \sqcup L), \\
&= \sum_{K \subseteq G, L \subseteq H} \mu(G \sqcup H, K \sqcup L) F(K \sqcup L), \\
&= \sum_{K \subseteq G, L \subseteq H} \mu(G \sqcup H - K \sqcup L, \emptyset) F(K \sqcup L).
\end{aligned}
$$

and we have also:

$$f(G) = \sum_{K \subseteq G} \mu(G-K,\emptyset)F(K)$$

and:

$$f(H) = \sum_{L \subseteq H} \mu(H-L,\emptyset)F(L).$$

Now, again, let us suppose that f is an additive function, then F is also an additive function:

$$
\begin{aligned}
f(G)f(H) &= \sum_{K \subseteq G} \mu(K-G,\emptyset)F(K) \sum_{L \subseteq H} \mu(H-L,\emptyset)F(L), \\
&= \sum_{K \subseteq G, L \subseteq H} \mu(K-G,\emptyset)\mu(H-L,\emptyset)F(K)F(L).
\end{aligned}
$$

We have that $K \cap L = \emptyset, \forall(K,L)$ such that $K \subseteq G$ and $L \subseteq H$ so that $F(K)F(L) = F(K \sqcup L)$.

$$
\begin{aligned}
f(G)f(H) &= \sum_{K \subseteq G} \mu(K-G,\emptyset)F(K) \sum_{L \subseteq H} \mu(H-L,\emptyset)F(L), \\
&= \sum_{K \subseteq G, L \subseteq H} \mu(K-G,\emptyset)\mu(H-L,\emptyset)F(K \sqcup L).
\end{aligned}
$$

By identifying $f(GH)$ and $f(G)f(H)$ we find again that:

$$\mu(K-G,\emptyset)\mu(H-L,\emptyset) = \mu(K-G \sqcup H-L,\emptyset).$$

Since μ is an additive function, we can easily compute its values on any sets.

Let us suppose that $B = A \sqcup \{x_1\} \sqcup \ldots \sqcup \{x_N\}$ thus $\mu(B-A,\emptyset) = \prod_{i=1,\ldots,N} \mu(x_i,\emptyset)$ and since for a singleton $\{x\}$, $\mu(\{x\},\emptyset) = -1$, we have

$$\mu(B-A,\emptyset) = (-1)^N = (-1)^{|B-A|}.$$

4) Let us evaluate the number $C(n,k,q)$ of subspaces of dimension k in a n-vector space over a finite field $\mathbb{K}$ with $\#\mathbb{K} = q$.

Let us prove that:

$$C(n+1,k+1,q) = q^{n-k}C(n,k,q) + C(n,k+1,q).$$

Any subspace of dimension $k+1$, $F(k+1)$, in a $n+1$-vector space E is either included in the Hyperplane $H(e_1,\ldots,e_n)$ spanned by the first n vector of the basis of E or not included in H.

The first case is described by $C(n,k+1,q)$ subspaces.

If $F(k+1)$ is not included in $H(e_1,\ldots,e_n)$, then $F(k)$ intersects $H(e_1,\ldots,e_n)$ in $F'(k)$, a k-linear subspace of $H(e_1,\ldots,e_n)$.

We show that the amount of $(k+1)$-linear subspace of E whose intersection with $H(e_1,\ldots,e_n)$ results in a given $F'(k)$ is q^{n-k}:

$F(k+1)$ is uniquely defined by a set of $(n+1-(k+1))$ equations:

$$\lambda_1^1 x_1 + \ldots + \lambda_{n+1}^1 x_{n+1} = 0,$$

$$\cdots$$

$$\lambda_1^{n-k} x_1 + \ldots + \lambda_{n+1}^{n-k} x_{n+1} = 0.$$

With: $\lambda_i^j \in \mathbb{K}, i = 1 \ldots n+1, j = 1 \ldots n-k$.

The intersection $F'(k)$ is defined by $n-k$ equations by making $x_{n+1} = 0$:

$$\lambda_1^1 x_1 + \ldots + \lambda_n^1 x_n = 0,$$

$$\cdots$$

$$\lambda_1^{n-k} x_1 + \ldots + \lambda_n^{n-k} x_n = 0.$$

Thus, for every k-subspace in $H = H(e_1,\ldots,e_n)=\text{span}(e_1,\ldots,e_n)$, the parameters $\lambda_{n+1}^j, (j = 1 \ldots n-k)$ define a $k+1$ subspace of E that intersects with H in $F'(k)$, so there are q^{n-k} case and thus[12]:

---

[12]In fact, we have: $C(n,k,q) = \dfrac{(q^n-1)\times\ldots\times(q-1)}{[(q^k-1)\times\ldots\times(q-1)][(q^{n-k}-1)\times\ldots\times(q-1)]}$

$$
\begin{aligned}
C(n+1,k+1,q) \;=\;& \text{amount of (k+1)-subspaces in H} \\
& +q^{n-k} \times \text{amount of k-subspaces in H,} \\
\;=\;& q^{n-k}C(n,k,q)+C(n,k+1,q).
\end{aligned}
$$

Now, let us prove that:

$$
\sum_{k=0}^{n} q^{\frac{k(k-1)}{2}} C(n,k,q) \times t^k = (1+t)(1+qt) \times \ldots \times (1+q^{n-1}t).
$$

If we define:

$$
f_{n,q}(t) = (1+t)(1+qt) \times \ldots \times (1+q^{n-1}t) = \sum_{k=0}^{n} a_{n,k}t^k.
$$

We have:

$$
f_{n+1,q}(t) = (1+q^n t)f_{n,k}(t) = (1+q^n t)\sum_{k=0}^{n} a_{n,k}t^k.
$$

So:

$$
(1+q^n t)\sum_{k=0}^{n} a_{n,k}t^k = \sum_{k=0}^{n+1} a_{n+1,k}t^k.
$$

Therefore:

$$
a_{n+1,k+1} = a_{n,k+1} + q^n a_{n,k}.
$$

If we put $a_{n,k} = q^{k(k-1)/2}c_{n,k}$, then:

$$
\begin{aligned}
a_{n,k+1}+q^n a_{n,k} \;=\;& q^{k(k+1)/2}c_{n,k+1}+q^n q^{k(k-1)/2}c_{n,k}, \\
\;=\;& q^{k(k+1)/2}(c_{n,k+1}+q^{n-k}c_{n,k}).
\end{aligned}
$$

So: $c_{n+1,k+1} = q^{n-k} c_{n,k} + c_{n,k+1}$ and thus: $c_{n,k} = C(n,k)$.

We start now to compute the Möbius function:

We know that μ is uniquely defined by the system:

$$\sum_{B \subseteq C \subseteq A} \mu(C,B) = \delta_{A,B}$$

and by the same argument as shown in a), b) and c) we have $\mu(C,B) = \mu(C/B,0)$ so that we only have to care about the values $\mu(C,0)$.

By making $t = -1$ in $f_{n,q}(t)$, we have $f_{n,q}(-1) = 0$ and:

$$\sum_{k=0}^{n} q^{\frac{k(k-1)}{2}} C(n,k,q) \times (-1)^k = 0.$$

We read clearly:

$$\sum_{0 \subseteq F \subseteq E} \mu(F,0) \times C(n,k,q) = 0, \dim(F) = k, \dim(E) = n.$$

So that:

$$\mu(C,B) = \mu(C/B,0) = (-1)^k \times q^{\frac{\dim(C/B)(\dim(C/B)-1)}{2}},$$

$$\mu(C,B) = (-1)^k \times q^{\frac{\dim(C)-\dim(B)(\dim(C)-\dim(B)1-1)}{2}}.$$

11/828 ¶¶

Subject: Use of the Möbius functions in $(\mathbb{N}, |)$ for calculating special number functions

Express the following functions by means of the Möbius function in $(\mathbb{N}, |)$:

a) The *Euler function* $\varphi(n)$ that is equal to the amount of positive numbers that are inferiors and primes to n;

b) The amount $P(n,q)$ of irreducible polynomials of degree n whose coefficients belongs to the finite field F_q and whose upper coefficient is 1;

c) The limit $C(N)/N^2, N \to \infty$ where $C(N)$ is the amount of irreducible fractions of the form $p/q, 1 \le (p,q) \le N$.

SOLUTION:

a) Let us compute $F(n) = \sum_{d|n} \varphi(d)$.

For every $d|n$, the sets $S_d = \{\frac{n}{d} \times m, (m|d) = 1, m < d\}$ are disjoint because if $x \in (S_d \cap S_{d'})$, then:

$$x = \frac{n}{d'} \times m' = \frac{n}{d} \times m.$$

If $d \ne d'$, then, we can suppose, for example that $d > d'$ and then we can find a $k > 1$ such that $kd' = d, km' = m$: indeed, from $dm' = d'm, d > d'$, we get that necessarily, $d'|d$ and $m'|m$, then we may write $k = \frac{d}{d'} = \frac{m}{m'}, k \in \mathbb{N}$.

But this is impossible because that would mean: $(d,m) \ne 1$, so we must have $S_d \cap S_{d'} = \emptyset$ for $d \ne d'$.

If $x \in \mathbb{N}, 1 \leq x \leq n$, then we can find a d such that $x \in S_d$, let s=g.c.d (n,x)
then: $x = sm, n = sd$ and $(m,d) = 1, m < d$ so that $x = (n/d) \times m \in S_d$.
    So we have:

$$[|1,n|] = \bigsqcup_{d|n} S_d$$

and:

$$n = \#[|1,n|] = \sum_{d|n} |S_d| = \sum_{d|n} \varphi(d).$$

We can invert $F(n) = n = \sum_{d|n} \varphi(d)$ by the inversion formula and we
get:

$$\varphi(n) = \sum_{d|n} d \times \mu(n,d) = \sum_{d|n} d \times \mu(n/d, 1)$$

or, using convolution:

$$\varphi = x * \mu.$$

b) To every polynomial P in $\mathbb{F}_q[X]$ of degree d and of upper coefficient
1, we associate the sequence:

$$f_P(X) = 1 + X^d + X^{2d} + \dots.$$

Let us show that

$$\prod_P f_P(X) = 1 + qX + q^2 X^2 + \dots.$$

We note E_d the space of irreducible polynomials of degree d with upper
coefficient $= 1$ and G_n the space of polynomials of degree n with upper
coefficient $=1$.
    Obviously, if $P \in E_d$ and $Q \in E_{d'}$, then PQ is a polynomial of degree
$d + d'$ with upper coefficient $= 1$.

Conversely, let us suppose that $P \in G_n$ decomposes in[13]:

$$P = \prod_{i=1,\ldots,\alpha_n} P_i^{a_i}, P_i \in E_{d_i}.$$

We know that this decomposition is unique.
We define the application ψ:

$$\begin{aligned} G_n & \to \langle E_i \rangle, \\ \psi(P) & = (a_i, d_i)_{i=1,\ldots,\alpha_n}. \end{aligned}$$

ψ is an injection, then it is a bijection from G_n into $\psi(G_n)$:
so:

$$|\psi(G_n)| = |G_n| = q^n$$

then:

$$\prod_P f_P(X) = \sum_n |\psi(G_n)| X^n = \sum q^n X^n.$$

Alternatively, we know that:

$$f_P(X) = \frac{1}{1 - X^d},$$

$$\sum q^n X^n = \frac{1}{1 - qX}.$$

So that:

$$\prod_d \left(\frac{1}{1 - X^d}\right)^{P(d,q)} = \frac{1}{1 - qX}.$$

By taking the log-derivate of the two members:

---

[13]The irreducible decomposition of P is necessarily done by polynoms with upper coefficients$= 1$ (except the polynom of degree $=0$)

$$\sum_d \frac{P(d,q)\frac{\partial}{\partial X}(\frac{1}{1-X^d}) \times (\frac{1}{1-X^d})^{P(d,q)-1}}{(\frac{1}{1-X^d})^{P(d,q)}} = \frac{q(\frac{1}{1-qX})^2}{\frac{1}{1-qX}},$$

$$\sum_d \frac{P(d,q) \times d \times (\frac{1}{1-X^d})^2 \times (\frac{1}{1-X^d})^{P(d,q)-1}}{(\frac{1}{1-X^d})^{P(d,q)}} = \frac{q(\frac{1}{1-qX})^2}{\frac{1}{1-qX}},$$

$$\sum_d P(d,q) \times d \times \frac{1}{1-X^d} = q\frac{1}{1-qX},$$

$$\sum_d P(d,q) \times d \times \sum X^{kd} = q\sum^n q^n X^n.$$

We must therefore have:

$$\sum_{d|n} dP(d,q) = (xP(x,q)*1)(n) = q^{n+1}.$$

The Möbius inversion formula tells us that:

$$nP(n,q) = \sum_{d|n} \mu(n,d)q^{d+1} = \sum_{d|n} \mu(n/d,1)q^{d+1},$$

$$P(n,q) = \frac{1}{n}\sum_{d|n} \mu(n/d,1)q^{d+1}.$$

c) Let us prove that

$$C(N) = 2 \sum_{1 \le k \le N} \varphi(k) - 1,$$

where $C(N)$ is the amount of irreducible fractions $\frac{p}{q}/(p,q) \le N$ [14].
For every $p \le N$, there is exactly $\varphi(p)$ numbers $q \le p$ so that $\frac{p}{q}$ is
irreducible and for every $q \le N$ there is exactly $\varphi(q)$ numbers $p \le q$ so

---

[14]If X is an ordered set and $(a,b,c) \in X^3$, we note $(a,b) \le c$ for $a \le c$ and $b \le c$

that $\frac{p}{q}$ is irreducible. The two cases are disjoint except for one case where: $p = q = 1$. So we must have:

$$
\begin{aligned}
C(N) &= \sum_{1 \leq p \leq N} \varphi(p) + \sum_{1 \leq q \leq N} \varphi(q) - 1, \\
&= 2 \sum_{1 \leq k \leq N} \varphi(k) - 1.
\end{aligned}
$$

We know from a) that $\varphi(n) = \sum_{d|n} d \times \mu(n/d, 1) = \sum_{d|n} \frac{n}{d} \times \mu(d, 1)$,

$$
C(N) = 2 \sum_{1 \leq k \leq N} \sum_{d|k} \frac{k}{d} \times \mu(d, 1) - 1,
$$

$$
C(N) = 2 \sum_{1 \leq d \leq N} \sum_{1 \leq k \leq N/d|k} \frac{k}{d} \times \mu(d, 1) - 1.
$$

We have $k = \alpha d, \alpha = 1, \dots, [\frac{N}{d}]$ so that:

$$
C(N) = 2 \sum_{1 \leq d \leq N} \mu(d, 1) \sum_{1 \leq \alpha \leq [\frac{N}{d}]} \alpha - 1,
$$

$$
C(N) = \sum_{1 \leq d \leq N} [\frac{N}{d}]([\frac{N}{d}] + 1) \times \mu(d, 1) - 1.
$$

So we have:

$$
C(N)/N^2 = \sum_{1 \leq d \leq N} ([\frac{N}{d}]^2/N^2 \times \mu(d, 1) + [\frac{N}{d}]/N^2 \times \mu(d, 1)) - 1/N^2,
$$

$$
\lim_{N \to \infty} C(N)/N^2 = \sum_d \mu(d, 1)/d^2 < \infty.
$$

We can compute $\sum_d \frac{\mu(d,1)}{d^2}$:

$$\sum_{d\geq 1}\frac{\mu(d,1)}{d^2}\times\sum_n\frac{1}{n^2} = \sum_{d\geq 1, n\geq 1}\frac{\mu(d,1)}{(dn)^2},$$

$$= \sum_{d\geq 1, d|N}\frac{\mu(d,1)}{N^2},$$

$$= \mu(1,1)/1^2 + \sum_N\frac{1}{N^2}\sum_{d|N}\mu(d,1),$$

$$= 1,$$

because of the fundamental property of the Möbius functions[15].
Now:

$$\lim_{N\to\infty}C(N)/N^2 = \sum_d\frac{\mu(d,1)}{d^2} = (\sum_n\frac{1}{n^2})^{-1} = \frac{6}{\pi^2}. \quad [16]$$

---

[15]See Exercise 8

[16]This is a classical result of elementary analysis, however this will be proved in exercises
536-c or 729

12/828 ¶

Subject: Use of the Möbius functions in $(\mathbb{N}, |)$ for building a family of polynomials

We consider μ, the Möbius function, we can build the family Φ_n by:

$$\Phi_n(x) = \prod_{d|n}(x^{n/d} - 1)^{\mu(d,1)}.$$

Prove that:

a) Φ_n is a polynomial of degree $\varphi(n)$ (φ being The Euler function) with entire coefficients.

b) $\prod_{d|n}\Phi_d(x) = x^n - 1$.

c) The Φ_n are irreducible and primes between each other over the field $\mathbb{Q}$ of rational numbers.

SOLUTION:

We note $\mu(d)$ the Möbius function $\mu(d,1)$ over $\mathbb{N}$ partially ordered by divisibility as defined as in Exercise 10)b so that Φ_n is the polynomial defined by $\Phi_n(X) = \prod_{d|n}(X^{n/d} - 1)^{\mu(d)}$.

We first investigate two methods: a method using the Möbius inversion formula and recursion (i), and a method using direct calculation of Φ_n (ii).

i)

If f and g are being defined by $f(X) = \ln(X^n - 1)$ and $g(X) = 1$.

Then we have, by noting $*$ the usual convolution product:

$$\ln \Phi_n(X) = \sum_{d|n} \ln(X^{n/d} - 1)\mu(d) = (f * \mu)(X).$$

So that, by Möbius inversion formula, we have:

$$f(X) = (\ln(\Phi_d) * g)(X) = \sum_{d|n} \ln(\Phi_d)(X).$$

What finally lead to:

$$\prod_{d|n} \Phi_d(X) = X^n - 1.$$

From this, we get the following recursive formula that allows us to calculate the $\Phi_n(X)$:

$$\Phi_{n+1}(X) = \frac{X^{n+1} - 1}{\prod_{d|n+1, d \neq n+1} \Phi_d(X)}.$$

By recursion we prove that the Φ_n are polynomials of $\mathbb{Q}[X]$ of degree $\varphi(n)$ whose roots are the n-root of unity (in $\mathbb{C}$). This is true for $n = 1$ now let us assume this is true for all $k \leq n$. Then we can see easily from the recursion formula that all the roots of the $\Phi_d(X)$ for $d|n+1, d \neq n+1$ are also roots of the polynomial $X^{n+1} - 1$: indeed a root of such a $\Phi_d(X)$ is a d-root of unity ζ so that $\zeta^d = 1$ and hence $\zeta^{n+1} - 1 = 0$. This shows that $\prod_{d|n=1, d \neq n+1} \Phi_d(X)$ divide $X^{n+1} - 1$ because all the roots of the Φ_d's are all distinct by definition since they are d-roots of unity. The degree of $\Phi_n(X)$ is therefore:

$$deg(X^{n+1} - 1) - deg(\prod_{d|n=1, d \neq n+1} \Phi_d(X)) = n + 1 - \sum_{i=1}^{i=n} \varphi(i) = \varphi(n+1). \quad [17]$$

ii)

---

[17] See properties of the Euler function φ

Without using this recursion process, we could try to get formulas for the polynomials $\Phi_d(X)$ using the properties of the Möbius function.

If $d = p$, a prime, then $\Phi_p(X) = (X^p - 1)^{\mu(1)}(X - 1)^{\mu(p)} = (X^p - 1)/(X - 1) = 1 + X + X^2 + \ldots + X^{p-1} \in \mathbb{Q}[X]$ and $deg(\Phi_p) = p - 1 = \varphi(p)$.

If $r > 0$, then $\Phi_{p^r}(X) = \prod_{d|p^r}(X^{p^r/d} - 1)^{\mu(d)} = (X^{p^r} - 1)^{\mu(1)}(X^{p^{r-1}} - 1)^{\mu(p)} = (X^{p^r} - 1)(X^{p^{r-1}} - 1)^{-1}$ (since $\mu(n) = 0$ if $p^2|n$, p being a prime), so that $\Phi_{p^r}(X) = 1 + X^{p^{r-1}} + X^1 + \ldots + X^{p-1}{}^{p^{r-1}} = \Phi_p(X^{p^{r-1}})$. This shows that Φ_{p^r} is a polynomial from $\mathbb{Q}[X]$ of degree $(p - 1)p^{r-1} = \varphi(p^r)$.

Next we consider the more general case $n = p^r q$, where p is a prime and $(p,q) = 1$ (e.g. p and q has no common non-trivial divisors), then $\Phi_n(X) = \prod_{d|p^r q}(X^{p^r q/d} - 1)^{\mu(d)}$ but the divisors of $p^r q$ can be written equivalently $p^i d$, where $0 \le i \le r$ and $d|q$, from this divisors, if $i > 1$, then $\mu(p^i d) = 0$ so we are only consider the case where $i = 0$ or $i = 1$ and we get:

$$\Phi_n(X) = \prod_{d|q}(X^{p^r q/d} - 1)^{\mu(d)}(X^{p^{r-1}q/d} - 1)^{\mu(pd)}.$$

But since the multiplicativity of μ imply that $\mu(pd) = \mu(p)\mu(d) = -\mu(d)$, we can rewrite this product as:

$$\Phi_n(X) = \prod_{d|q}(X^{p^r q/d} - 1)^{\mu(d)} / \prod_{d|q}(X^{p^{r-1}q/d} - 1)^{\mu(d)}.$$

That is to say:

$$\Phi_n(X) = \frac{\Phi_q(X^{p^r})}{\Phi_q(X^{p^{r-1}})}.$$

From here, we deduce that if Φ_q is a polynomial of degree $f(q)$, then $\Phi_{p^r q}$ is a polynomial of degree $p^r f(q) - p^{r-1} f(q) = p(p^{r-1})f(q)$. Indeed, if α is a root of $\Phi_q(X^{p^{r-1}})$ then $\beta = \alpha^{p^{r-1}}$ is a root of Φ_q. We see easily that β^p will be also a root of Φ_q because $\Phi_q(\beta) = 0$ is equivalent to the giving of a number d such that $d|q$ and $\beta^{q/d} = 1$, $\mu(d) \ne 0$, this implies that $(\beta^p)^{q/d} = 1$ and this means that β^p is a root of Φ_q. So that we may conclude that $\Phi_q(\alpha^{p^r}) = 0$ and that the roots of the denominator are also roots of the numerator.

If $n = p_1^{r_1} \ldots p_s^{r_s}$, then we apply this process to $p_1^{r_1}$, $p_1^{r_1} p_2^{r_2}$, $\ldots$ untill we reach $n = p_1^{r_1} \ldots p_s^{r_s}$ and we get that $\Phi_{p_1^{r_1} p_2^{r_2}}$ is a polynomial of degree

$$p_2(p_2^{r_2-1})\varphi(p_1^{r_1}) = p_2(p_2^{r_2-1})p_1(p_2^{r_1-1}) \text{ that is to say of degree } \varphi(p_1^{r_1} p_2^{r_2})$$

etc... We finally get that Φ_n is a polynomial of degree $p_1(p_2^{r_1-1}) \ldots p_s(p_s^{r_s-1}) = \varphi(n)$.

We can now show that if ζ is a n-root of unity then $\Phi_n(\zeta) = 0$. This is rather immediate from the definition of Φ_n. Indeed,

$$\Phi_n(\zeta) = \prod_{d|n}(\zeta^{n/d} - 1)^{\mu(d)} = (\zeta^n - 1)^{\mu(1)} \times \prod_{d|n, d>1}(\zeta^{n/d} - 1)^{\mu(d)}.$$

So that:

$$\Phi_n(\zeta) = 0.$$

Therefore, as there are $\varphi(n)$ n-roots of unity they form the roots of Φ_n.

To prove now, without the Möbius inversion formula , that $X^n - 1 = \prod_{d|n} \Phi_d(X)$ is straightforward: indeed we can partition the roots of unity in the union of the d-roots of unity for numbers d such that $d|n$:

$$\{\text{roots of } X^n - 1 = 0\} = \bigsqcup_{d|n} \{\text{ d-roots of unity }\}$$

so that:

$$\prod_{\zeta \in \{\text{roots of } x^n - 1 = 0\}} (X - \zeta) = \prod_{d|n} \prod_{\zeta \in \{\text{ d-roots of unity }\}} (X - \zeta).$$

The Φ_d's having all leading terms $=1$, this implies the result.

Whatever methods we have used (i or ii) we need now to prove that the Φ_n are in $\mathbb{Z}[X]$ for this, we can use the following theorem: If $(f,g) \in \mathbb{Z}[X]$ and g has a leading term$=1$, then we can find uniquely two polynomials $(q,r) \in \mathbb{Z}[X]$ such that $deg(r) < deg(g)$ and $f = gq + r$ (Euclidean division principle in $\mathbb{Z}[X]$) and we can conclude by recursion but we can use the more general result form the Gauss lemma in $\mathbb{Q}[X]$: let $f \in \mathbb{Q}[X]$.

If $a \in \mathbb{Q}$, then, if we have $a = p^r \frac{m}{n}$ with $\Delta(m, n) = 1, \Delta(m, p) = 1, \Delta(n, p) = 1$, we note $ord_p(a) = r$ (that is equivalent to the definition $||a||_p = p^{-ord_p(a)}$, where $||.||_p$ is the p-valuation over $\mathbb{Q}$ introduced in Exercise 37). Then if $f(X) = \sum_{i \leq N} a_i X^i$, $ord_p(f)$ is define as ∞ if $f = 0$ and $min\{ord_p(a_i), a_i \neq 0\}$ if $f \neq 0$. Next we define $cont(f)$ as being equal to the product $\prod p^{ord_p(f)}$, the sum being taken over all prime numbers p such that $ord_p(f) \neq 0$ (there are only a finite amount of such p's).

For example, if $f(X) = 2 + \frac{1}{9}X + \frac{1}{3}X^2 + 8X^3$, then $ord_1(f) = 1$, $ord_2(f) = 1$, $ord_3(f) = min(-2, -1) = -2$ and the others ord_p are $=0$, we therefore get: $cont(f) = 2 \times 3^{-2} = 2/9$.

If $f(X) = \sum_{p \leq n, p \in \pi} p^{-n_p} X^p$, then $cont(f) = \prod_{p \leq n, p \in \pi} p^{-n_p}$.

We see also easily that: $f \in \mathbb{Z}[X] \Leftrightarrow cont(f) = 1$.

The Gauss Lemma tells us that for two polynomials (f, g) in $\mathbb{Q}[X]$, $cont(f) \times cont(g) = cont(fg)$.

We apply this here with $f(X) = \Phi_n(X)$ and $g(X) = \prod_{d|n, d \neq n} \Phi_d(X)$. If we suppose by recursion that the Φ_m's are all in $\mathbb{Z}[X]$ for $m < n$ then $cont(g) = 1$ and as $cont(X^n - 1) = 1$, we must have $cont(\Phi_n) = 1$ what imply that $\Phi_n \in \mathbb{Z}[X]$.

The last part is to show that the Φ_d's are irreducible over $\mathbb{Q}$. The fact that they are primes together is almost immediate since they cannot have, by definition of n-primitive roots of unity, any root in common.

To show that they are irreducible over $\mathbb{Q}$, let us proceed as follows:

let us suppose that $f(X)$ is an irreducible polynomial in $\mathbb{Q}$ (f being a new polynomial) and that $f(\zeta) = 0$ for a n-primitive root of unity ζ. Then we will have $X^n - 1 = f(X).h(X)$ in $\mathbb{Q}[X]$. We can use again Gauss Lemma to see that $cont(f)cont(h) = 1$. Since f is irreducible it must have leading term $= 1$ and then so must be h so this implies $cont(f) = 1$ and $cont(h) = 1$ that is to say $f(X)$ and $h(X)$ are in $\mathbb{Z}[X]$. If we can prove that for any prime number $p \leq n$ such that p does not divide n then $f(\zeta^p) = 0$ then we will be done since the ζ^p's generated like that will be the $\varphi(n)$ n-primitive roots of unity.

To do this, let us suppose on the contrary that for a prime p, ζ^p is not a root of f then we must have $h(\zeta^p) = 0$ and hence ζ will be a root of $h(X^p)$, then again we can write $h(X^p) = f(X)g(X)$ with $g(X) \in \mathbb{Z}[X]$ (where f and g are some new polynomials).

Let $h(X) = \sum_{i=1}^{m} a_i X^i, a_i \in \mathbb{Z}$, then:

$$h(X)^p = (\sum_{i=1}^{m} a_i X^i)^p = \sum_{k_0 + \ldots + k_m = n} \binom{p}{k_0, \ldots, k_m} (a_0^{k_1} (a_1 X)^{k_1} \times \ldots (a_m X^m)^{k_m} \; [18].$$

We know that $p | \binom{p}{k_0, \ldots, k_m}$ when these coefficients are $\neq 1$. So that, working modulo p, the only cases to be considered are $k_0 = p, k_1 = \ldots = k_m = 0, \ldots k_0 = \ldots = k_{m-1} = 0, k_m = p$. So we get $h(X)^p = a_0^p + (a_1 X)^p + \ldots (a_m X^m)^p (mod\, p)$ and by Fermat's little theorem, we have:

$$a_0^p = a_0 (mod\, p), \ldots, (a_m X^m)^p = a_m X^m (mod\, p),$$

so: $h(X)^p = a_0 + a_1 X + \ldots + a_m X^m = h(X)(mod\, p)$.
This leads to $h(X)^p = f(X)g(X)(mod\, p)$.
Let $\overline{f}(X)$ be the polynomial from $\mathbb{Z}/p\mathbb{Z}[X]$ defined by $\overline{f}(X) = f(X)(mod\, p)$ and $\overline{g}(X)$ be all the same $g(X)(mod\, p)$ then we will have:
$\overline{f}(X)\overline{g}(X) = \overline{h}(X)^p$.
This implies that $\overline{f}$ and $\overline{g}$ are not prime of each other.
From $X^n - \overline{1} = \overline{f}(X)\overline{g}(X)$, we deduce that $X^n - \overline{1}$ has at least one root of order ≥ 2.
This leads to $X^n - \overline{1} = (X - \overline{a})^2 \overline{Q}(X), \overline{a} \neq 0$.
But by deriving the two sides: $nX^{n-1} = 2(X - \overline{a})\overline{Q}(X) + (x - \overline{a})^2 \overline{Q}'(X)$ so that we finally get $n\overline{a}^{n-1} = 0, \overline{a} \neq 0$, which is impossible modulo p.
Then we conclude that all irreducible polynomials f such that $f(\zeta) = 0$, ζ being a n-primitive root of unity must have at least $\varphi(n)$ roots, what imply the irreducibility of the Φ_n's.
The Φ_n's are called the cyclotomic polynomials.

---

[18] $\binom{p}{k_0, \ldots, k_m} = \frac{p!}{k_1! \ldots k_m!}$ are the multinomial coefficients

13/828

Subject: Lexicographic order

Let us suppose that A is a well ordered set and that: $\forall \alpha \in A$ we can then define an ordered set $X_\alpha \neq \emptyset$. We define the *lexicographic* order on $X = \prod_\alpha X_\alpha$ by:

$$x \geq y \text{ if } x_{\alpha_0} \geq y_{\alpha_0}, \alpha_0 = \min\{\alpha, x_\alpha \neq y_\alpha\}.$$

Show that it is an order relation.

SOLUTION:

The first thing to check is that the lexicographical order is well-defined. Let $A(x,y)$ be the subset of A defined by $A(x,y) = \{\alpha \in A, x_\alpha \neq y_\alpha\}$. Then, because A is well ordered, $A(x,y)$ has a minimum element α_0 and this ensure us of the validity of the definition of the lexicographical order.

If $x \geq y$ and $y \geq x$. If $\min\{\alpha, x_\alpha \neq y_\alpha\} < \infty$:

$$x_{\alpha_0} \geq y_{\alpha_0}, \alpha_0 = \min\{\alpha, x_\alpha \neq y_\alpha\}.$$

And:

$$y_{\beta_0} \geq x_{\beta_0}, \beta_0 = \min\{\beta, x_\beta \neq y_\beta\}.$$

So that $\alpha_0 = \beta_0$ and:

$$(x_{\alpha_0} \geq y_{\alpha_0}, x_{\alpha_0} \leq y_{\alpha_0}) \quad \Longrightarrow \quad (x_{\alpha_0} = y_{\alpha_0}),$$
$$\Longrightarrow \quad \min\{\alpha, x_\alpha \neq y_\alpha\} \neq \alpha_0.$$

We must therefore have $x = y$ (antisymmetry).
If $x \geq y \geq z$:

$$x_{\alpha_0} \geq y_{\alpha_0}, y_{\beta_0} \geq z_{\beta_0},$$

$$\alpha_0 = \min\{\alpha, x_\alpha \neq y_\alpha\}, \beta_0 = \min\{\beta, y_\beta \neq z_\beta\}.$$

If $\alpha_0 \geq \beta_0$, then $x_{\beta_0} = y_{\beta_0} \geq z_{\beta_0}$.
If $\alpha_0 \leq \beta_0$, then $x_{\alpha_0} \geq y_{\alpha_0} = z_{\alpha_0}$.
So that, if $\gamma_0 = \min(\alpha_0, \beta_0)$, $x_{\gamma_0} \geq z_{\gamma_0}$.
We have $x_{\gamma_0} \neq z_{\gamma_0}$.
If $\gamma_0' < \gamma_0$ then we have:

$$x_{\gamma_0'} = y_{\gamma_0'} = z_{\gamma_0'}.$$

So that:

$$\gamma_0 = \min\{\gamma, x_\gamma \neq z_\gamma\}$$

and so, finally: $x \geq z$ (transitivity).
This is a total order because if: $x \neq y$ and $\alpha_0 = \min\{\alpha, x_\alpha \neq y_\alpha\} < \infty$
then:

$$x_{\alpha_0} < y_{\alpha_0} \text{ or } x_{\alpha_0} > y_{\alpha_0}.$$

The lexicographic order allows us to create a total order on a product of sets.

14/828

Subject: Isomorphism between $\mathbb{R}^n$ and a product of sets provided with a lexicographic order

We suppose that $\mathbb{R}^n$ is provided with an order $\preceq$ that satisfies the following conditions:

1. $\vec{x_1} \succeq \vec{y_1} \wedge \vec{x_2} \succeq \vec{y_2} \implies \vec{x_1} + \vec{x_2} \succeq \vec{y_1} + \vec{y_2}$,

2. $\vec{x} \succeq \vec{y} \wedge \lambda \geq 0 \implies \lambda \vec{x} \succeq \lambda \vec{y} \, \forall \lambda \in \mathbb{R}$,

3. $\vec{x} \succeq \vec{y} \wedge \vec{y} \succeq \vec{x} \implies \vec{x} = \vec{y}$.

Show that $\mathbb{R}^n$ provided with this order is isomorph to the product $\prod_{i=1}^{i=n} \mathbb{R}$ of n copies[19] of $\mathbb{R}$ provided with a lexicographic order .

SOLUTION:

We consider the sets O^+ and O^- of, respectively, positives and negatives points of $\mathbb{R}^n$ defined as follows : if we call $\mathscr{O} = \{\mathscr{V}(\vec{x})\}$ the set of neighborhoods[20] of $\vec{x}$, then:

$$O^+ = \{\vec{x} \in \mathbb{R}^n, \exists \mathscr{V}(\vec{x}) \in \mathscr{O} / \forall \vec{x'} \in \mathscr{V}(\vec{x}), \vec{x'} \succeq \vec{0}\},$$

---

[19]Each copy of $\mathbb{R}$ being provided with the usual order $\leq$
[20]The topology used here being the Euclidean Topology of $\mathbb{R}^n$

$$O^- = \{\vec{x} \in \mathbb{R}^n, \exists \mathcal{V}(\vec{x}) \in \mathcal{O} / \forall \vec{x'} \in \mathcal{V}(\vec{x}), \vec{x'} \preceq \vec{0}\}.$$

We also define the set Γ by:

$$\Gamma = \overline{O^+} \cap \overline{O^-}.$$

We show that Γ is an hyperplane.

O^+ and O^- are non empty convex sets:

If $(x,y) \in (O^+)^2$, $(a,b) \in (\mathbb{R}^+)^2$, then, $\forall \vec{x'} \in \mathcal{V}(\vec{x}), \forall \vec{y'} \in \mathcal{V}(\vec{y})$: $a\vec{x'} + b\vec{y'} \succeq a \times \vec{0} + b \times \vec{0} = \vec{0}$ by using the properties 1) and 2) of the order on $\mathbb{R}^n$.

$a\mathcal{V}(\vec{x}) + b\mathcal{V}(\vec{y})$ is a neighborhood of $\mathcal{V}(a\vec{x} + b\vec{y})$ so that O^+ is a convex set. We show that O^- is also convex using the same method.

If $\vec{x} \succeq \vec{0}$, then $-\vec{x} \succeq -\vec{x}$ and from the property 1) we have $\vec{x} - \vec{x} \succeq \vec{0} - \vec{x}$ so that $\vec{x} \succeq \vec{0} \implies -\vec{x} \preceq \vec{0}$. By using the same method, we have $\vec{x} \preceq \vec{0} \implies -\vec{x} \succeq \vec{0}$.

Obviously, now, we can choose a basis of $\mathbb{R}^n$ made of positive/negative vectors $(\vec{x_1^+}, \ldots, \vec{x_n^+}) / (\vec{x_1^-}, \ldots, \vec{x_n^-})$.

$\mathbb{R}^n$ is provided with a total order and, for all $\vec{x} \in \mathbb{R}^n$, either $\vec{x}$ or $-\vec{x}$ is positive.

If $\vec{x} = \sum a_i \vec{x_i^+} \succeq \vec{0}$, then:

$$\mathcal{V}_\varepsilon(\vec{x}) = \sum_{0 < a_i - \varepsilon_i \le a_i' \le a_i + \varepsilon_i} a_i' \vec{x_i^+}$$

is a neighborhood of $\vec{x}$ where all vectors are positives. This shows that O^+ is non-empty.

As $O^- = -O^+$, O^- is also non empty.

Now, we consider $\vec{u} \in \Gamma$. We my find sequences $\{\vec{u_i^+}\}_{i \ge 0}$ and $\{\vec{u_i^-}\}_{i \ge 0}$ such that:

$$\vec{u} = \lim_{i \to \infty} \vec{u_i^+}, \vec{u_i^+} \in O^+$$

and:

$$\vec{u} = \lim_{i \to \infty} \vec{u_i^-}, \vec{u_i^-} \in O^-.$$

If $\vec{v} \in \Gamma$, we define in the same way the sequences $\{\vec{v_i^+}\}_{i \geq 0}$ and $\{\vec{v_i^-}\}_{i \geq 0}$. Then we have:

$$\vec{u} + \vec{v} = \lim_{i \to \infty}(\vec{u_i^+} + \vec{v_i^+}), (\vec{u_i^+} + \vec{v_i^+}) \in O^+,$$

$$= \lim_{i \to \infty}(\vec{u_i^-} + \vec{v_i^-}), (\vec{u_i^-} + \vec{v_i^-}) \in O^-.$$

So that $\vec{u} + \vec{v} \in \gamma$.
If $\lambda > 0$: then $\lambda \vec{u} \in \Gamma$ because of the convexity of O^+ and O^- [21].
If $\lambda < 0$:

$$\lambda \vec{u} = \lim_{i \to \infty} \lambda \vec{u_i^-}, \lambda \vec{u_i^-} \in O^+,$$

$$= \lim_{i \to \infty} \lambda \vec{u_i^+}, \lambda \vec{u_i}+ \in O^-.$$

So that, $\forall \lambda \in \mathbb{R}, \lambda \vec{u} \in \Gamma$.
This shows that Γ is a sub vector space of $\mathbb{R}^n$.
We consider $\vec{x^+} \in O^+$ and $\vec{x^-} \in O^-$, we define $[\vec{x^+}, \vec{x^-}]$ as:

$$[\vec{x^+}, \vec{x^-}] = \{a\vec{x^+} + (1-a)\vec{x^-}, 1 \leq a \leq 0\}.$$

We show that $\exists \vec{x}/\vec{x} \in [\vec{x^+}, \vec{x^-}] \cap \Gamma$:
We define λ^+ and λ^- as:

$$\lambda^+ = \inf\{\lambda \in [0,1]/\lambda\vec{x^+} + (1-\lambda)\vec{x^-} \in O^+\}[22].$$

And:

$$\lambda^- = \sup\{\lambda \in [0,1]/\lambda\vec{x^+} + (1-\lambda)\vec{x^-} \in O^-\}.$$

---

[21] One can also admits the result that the intersection of two convex bodies in $\mathbb{R}^n$ is a convex body or $\emptyset$

[22] The set if not empty since it contains at least $\lambda = 1$

We note also that $[\vec{x^+},\vec{x^-}] \subset O^+ \cup O^-$.

We must have $\lambda^+ = \lambda^- = \lambda_0$ since if there would be a λ' such that $\lambda^+ > \lambda' > \lambda^-$, we would have:

$$\lambda' \cap [\vec{x^+},\vec{x^-}] = \emptyset \qquad \Longrightarrow \qquad \lambda' \in [\vec{x^+},\vec{x^-}] - O^-,$$
$$\Longrightarrow \qquad \lambda'\vec{x^+} + (1-\lambda')\vec{x^-} \in O^+ \wedge (\lambda' < \lambda^+)$$

- which is impossible .

So that:

$$\lambda_0\vec{x^+} + (1-\lambda_0)\vec{x^-} \in \Gamma.$$

Γ is a sub vector space that separates $\mathbb{R}^n$ in two convex parts: O^+ and O^-, that is to say $\mathbb{R}^n\text{-}\Gamma$ has two connected (convex) components. If we use the geometrical form of the Hahn-Banach Theorem (TPFA, 3.1 Th 3), *we know that for any two disjoint convex sets U_1 and U_2 in $\mathbb{R}^n$ such that at least one the set is open, there exist an hyperplane H separating U_1 and U_2*. Now let us have a closer look at the sets O^+ and O^-. If p^+ is the set of positive points in $\mathbb{R}^n$, e.g. $p^+ = \{\vec{x}/\vec{x} \succeq 0, \vec{x} \in \mathbb{R}^n\}$, then $O^+ = \overset{\circ}{p}^{+}$[23]. With a similar definition for p^- we will have $O^- = \overset{\circ}{p}^-$.

Then $p^+ = -p^-, p^- = -p^+$ and hence $O^+ = -O^-, O^- = -O^+$.

We have also $O^+ \cap O^- = \{0\}$ (or $\emptyset$ depending if we define the sets as containing 0 or not).

Now we have also that $\overline{O^+} \setminus O^+ = \overline{O^-} \setminus O^-$, indeed, the elements of $\overline{O^+} \setminus O^+$ belong to either one or the other following two sets:

1) S_1^+: set of limits $\vec{x_\infty}$ of the sequences $\{\vec{x_i}\}$ of O^+ such that $\vec{x_\infty}$ is positive and $\forall \varepsilon > 0$, we can find an open ball $B(\vec{x_\infty},\varepsilon)$ of radius ε such that $B(\vec{x_\infty},\varepsilon) \cup O^- \neq \{0\}$.

2) S_2^+: set of limits $\vec{x_\infty}$ of the sequences $\{\vec{x_i}\}$ of O^+ such that $\vec{x_\infty}$ is negative.

If we use similar notations for O^- we may define S_1^- and S_2^-.

---

[23] $\overset{\circ}{A}$ is the interior of A

We check at once that $S_1^+ = S_2^-$ and $S_2^+ = S_1^-$ so that $\overline{O^+} \setminus O^+ = \overline{O^-} \setminus O^-$.

We know that $\mathbb{R}^n = p^+ \sqcup p^-$ [24] so that $\mathbb{R}^n = \overline{O^+} \cup \overline{O^-}$.

We have the following situation:

$$\mathbb{R}^n = (\setminus O^+ \sqcup (\overline{O^+} \setminus O^+)) \cup (O^- \sqcup (\overline{O^-} \setminus O^-))$$

and it turns out that:

$$\mathbb{R}^n = O^+ \sqcup O^- \sqcup ((\overline{O^-} \setminus O^-) \cap O^+) \sqcup ((\overline{O^+} \setminus O^+) \cap O^-) \sqcup ((\overline{O^-} \setminus O^-) \cap (\overline{O^+} \setminus O^+)).$$

Thus: $\mathbb{R}^n \setminus O^+ \cap \mathbb{R}^n \setminus O^- = (\overline{O^-} \setminus O^-) \cap (\overline{O^+} \setminus O^+) = \overline{O^+} \cap \overline{O^-}$.

Then, if H is an hyperplane separating O^+ and O^- (see Fig. 1), H must lie in $\mathbb{R}^n$ $(O^+ \cup O^-)$, that is to say: $H \in (\mathbb{R}^n \setminus O^+) \cap (\mathbb{R}^n \setminus O^-)$ this implies that H must lie in $\mathbb{R}^n \setminus - (O^+ \cup O^-) = \overline{O^+} \cap \overline{O^-} = \Gamma$, that is to say we must have $H = \Gamma$.

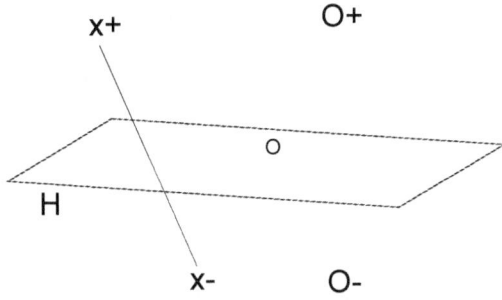

Figure 1: Separation between O^+ and O^- by H

By induction on n, we prove now that the order over $\mathbb{R}^n$ is isomorphic to a lexicographic order.

This is obviously true when $n = 1$, let us suppose that this property holds for all $k < n$.

We consider a basis of $\mathbb{R}^n$ formed by the set $\{\vec{e_n} \cup B(\Gamma)\}$, where $B(\Gamma)$ is a basis of Γ. Any vector $\vec{u} \in \mathbb{R}^n$ can be decomposed along Γ in:

---

[24]In what follows, by abuse of notation, we use the symbol $A \sqcup B$ if the intersection of A and B is either $\emptyset$ or $\vec{0}$

$$\vec{u} = \vec{u}_n + \vec{u}|\Gamma, \vec{v} = \vec{v}_n + \vec{v}|\Gamma.$$

We have two possibilities:

1) $\vec{u}_n = \vec{v}_n$, then $\min\{i/\vec{u}_i \neq \vec{v}_i\} < n$ and as Γ is provided with a lexicographic order isomorphic to the order in $\mathbb{R}^{n-1}$: $\vec{u} \succeq \vec{v} \Leftrightarrow \vec{u}|\Gamma \succeq \vec{v}|\Gamma$.

2) $\vec{u}_n \neq \vec{v}_n$, then $\min\{i/\vec{u}_i \neq \vec{v}_i\} = n$ and $\vec{u} - \vec{v} = (\vec{u}_n - \vec{v}_n) + \vec{z}_\Gamma, \vec{z}|_\Gamma \in \Gamma$ and $\vec{u} \succeq \vec{v} \Leftrightarrow \vec{u}_n \succeq \vec{v}_n$ (the component along Γ does not matter).

This shows by induction on n that the order in $\mathbb{R}^n$ is isomorphic to a lexicographic order.

Alternate solution[25].

We first start to prove two results for the order $\prec$:

a) If $\lambda < 0$ and $\vec{e}_1 \succ \vec{e}_2$, then we have $-\lambda > 0$ and this implies $-\lambda e_1 \succ -\lambda \vec{e}_2$. But as $\lambda \vec{e}_1 + \lambda \vec{e}_2 \succeq \lambda \vec{e}_1 + \lambda \vec{e}_2$ this implies also:

$$-\lambda \vec{e}_1 + (\lambda \vec{e}_1 + \lambda \vec{e}_2) \succeq -\lambda \vec{e}_2 + (\lambda \vec{e}_1 + \lambda \vec{e}_2)$$

that is to say:

$$\lambda \vec{e}_1 \preceq \lambda \vec{e}_2 \quad ^{26}.$$

b) If $\lambda_1 \geq \lambda_2$ then $\lambda_1 - \lambda_2 \geq 0$ and this implies:

$$\forall \vec{e}_1 \in \mathbb{R}^n, (\lambda_1 - \lambda_2)\vec{e}_2 \succeq 0$$

or:

$$\forall \vec{e}_1 \in \mathbb{R}^n, \lambda_1 \vec{e}_1 - \lambda_2 \vec{e}_2 \succeq 0.$$

By adding $\lambda_2 \vec{e}_2$ to both sides, we get:

$$\forall \vec{e}_1 \in \mathbb{R}^n, \lambda_1 \vec{e}_1 \succeq \lambda_2 \vec{e}_2.$$

Now we wish to prove the result by induction and we start with n=2.

---

[25]This solution needs to be completed .It is a good exercise for the reader to find where are the missing bits

[26]That is to say $\lambda \vec{e}_1 \prec \lambda \vec{e}_2$ since we cannot have $\lambda \vec{e}_1 = \lambda \vec{e}_2$

We let $U = \{\mu \in \mathbb{R}, \mu\vec{e}_1 - \vec{e}_2 \succ 0\}$. 1 is a least upper bound for U and since $(\forall \mu' \geq \mu), \mu \in U \Rightarrow \mu' \in U$, U is an interval of $\mathbb{R}$.

We let $\mu_1 = \inf(U)$, then:

case i) If $\mu_1 \notin U$, then we consider the function f:

$$\mathbb{R} \rightarrow \mathbb{R}^2,$$
$$\mu \rightarrow f(\mu) = \mu\vec{e}_1 - \vec{e}_2.$$

We provide $\mathbb{R}^2$ with the order topology from $\succ$.

Either f is sup-continuous in μ_1, or it is not.

The continuity of f in μ_1 mean that:

$$\forall(\vec{x},\vec{y}) \in \mathbb{R}^2 \times \mathbb{R}^2, \vec{x} \prec \mu_1\vec{e}_1 - \vec{e}_2 \prec \vec{y},$$
$$\exists \varepsilon > 0, 0 < \mu - \mu_1 < \varepsilon \Rightarrow 0 \prec \mu\vec{e}_1 - \vec{e}_2 \prec \vec{y}.$$

Sub case i-1) If f is not sup-continuous in μ_1, then we can find $a \in \mathbb{R}^2$ such that $\mu_1\vec{e}_1 - \vec{e}_2 \prec \vec{y}$. We can find a sequence $\{\varepsilon_i\}_{i=1}^{\infty}$, $\varepsilon_i \rightarrow 0$, $(\mu_1 + \varepsilon_i)\vec{e}_1 - \vec{e}_2 \succ \vec{y}$.

That means $\varepsilon_i\vec{e}_1 \succ \vec{y} + \vec{e}_2 - \mu_1\vec{e}_1 (\forall \varepsilon_i)$.

In that case, we can consider the new vectors:

$$\begin{aligned} \vec{e}_2{}' &= \vec{e}_1, \\ \vec{e}_1{}' &= \vec{y} + \vec{e}_2 - \mu_1\vec{e}_1. \end{aligned}$$

And we see that[27]:

$$\vec{e}_2{}' \succ \vec{e}_1{}' \succ 0,$$
$$\vec{e}_2{}' \succ \mathbb{R}\vec{e}_1{}'.$$

So that we may view $\prec$ as a lexicographical order on $\mathbb{R}^2$ provided by the basis $(\vec{e}_2{}', \vec{e}_1{}')$.

---

[27] We note $\vec{a} \succ \mathbb{R}\vec{b}$ if $\vec{a} \succ x\vec{b}, \forall x \in \mathbb{R}$

Sub case i-2) If we suppose that f is sup-continuous on μ_1, we have $\mu_1 \notin U$ so that:

$f(\mu_1) = \mu_1 \vec{e_1} - \vec{e_2} \prec 0$ and $\mu \to^+ \mu_1 \Rightarrow f(\mu) \to f(\mu_1)$ but $f(\mu_1) \prec 0 \prec f(\mu)(\forall \mu \in U)$ and this contradict the continuity of f.

Case ii) let us suppose that $\mu_1 \in U$. In that case: $(\forall \varepsilon > 0), f(\mu_1 - \varepsilon) = (\mu_1 - \varepsilon)\vec{e_1} - \vec{e_2} \prec 0$ or $\varepsilon \vec{e_1} \succ \mu_1 \vec{e_1} - \vec{e_2}$.

Then if we chose two new vectors by making:

$$\begin{aligned} \vec{e_2}' &= \vec{e_1}, \\ \vec{e_1}' &= \mu_1 \vec{e_1} - \vec{e_2}. \end{aligned}$$

We will have:

$$\begin{aligned} \vec{e_2}' &\succ \vec{e_1}' \succ 0, \\ \vec{e_2}' &\succ \mathbb{R}\vec{e_1}'. \end{aligned}$$

So that again, $\prec$ may be viewed as a lexicographical order provided by the basis $(\vec{e_2}', \vec{e_1}')$.

To conclude, in $\mathbb{R}^2$, on can always find a basis $(\vec{e_1}, \vec{e_2})$ such that:

$$\begin{aligned} \vec{e_2} &\succ \vec{e_1} \succ 0, \\ \vec{e_2} &\succ \mathbb{R}\vec{e_1}, \end{aligned}$$

e.g. a basis $(\vec{e_1}, \vec{e_2})$ where $\succ$ is a lexicographical order.

We would like to show that, in the general case, we can always find a basis $(\vec{e_1}, \ldots, \vec{e_n})$ of $\mathbb{R}^n$ such that:

$$\begin{aligned} \vec{e_n} &\succ \ldots \succ \vec{e_1} \succ \vec{0}, \\ \vec{e_n} &\succ \mathbb{R}\vec{e_{n-1}}, \vec{e_{n-1}} \succ \mathbb{R}\vec{e_{n-2}} \ldots \vec{e_2} \succ \mathbb{R}\vec{e_1}. \end{aligned}$$

By recursion, if we suppose that this is true for $\mathbb{R}^k$, $(\forall k < n)$.

We suppose we have found a basis such that: $\vec{e}_n \succ \vec{e}_{n-1} \succ \ldots \vec{e}_3 \succ \vec{e}_2 \succ \vec{e}_1 \succ 0$ and $\vec{e}_n \succ \mathbb{R}\vec{e}_{n-1}, \vec{e}_{n-1} \succ \mathbb{R}\vec{e}_{n-2} \ldots \vec{e}_2$, then, considering $\vec{e}_2$ and $\vec{e}_1$, we can find two new vectors, $\vec{e}_2'$ and $\vec{e}_1'$ such that $\vec{e}_2' \succ \vec{e}_1' \succ 0$ and $\vec{e}_2' \succ \mathbb{R}\vec{e}_1'$. In the process of building $\vec{e}_1'$ and $\vec{e}_2'$ we will always have $\vec{e}_2' \preceq \vec{e}_1$ so that $\vec{e}_n \succ \vec{e}_{n-1} \succ \ldots \vec{e}_3 \succ \vec{e}_2' \succ \vec{e}_1' \succ \vec{0}$ and since $\vec{e}_2 \succ \vec{e}_1 \succ \vec{e}_2'$, we have $\vec{e}_3 \succ \mathbb{R}\vec{e}_2 \succ \mathbb{R}\vec{e}_2'$. To conclude : $\vec{e}_n \succ \mathbb{R}\vec{e}_{n-1}, \vec{e}_{n-1} \succ \mathbb{R}\vec{e}_{n-2} \ldots \vec{e}_3 \succ \mathbb{R}\vec{e}_2', \vec{e}_2' \succ \mathbb{R}\vec{e}_1'$ and we have found a basis for $\mathbb{R}^n$ where $\prec$ may been viewed as a lexicographical order.

15/828 ¶¶

Subject: ω^1, the first uncountable ordinal

We say that two enumerable well-ordered sets A, B are equivalent ($A \sim B$) if there exists a (strictly) monotonous bijection between each other. Let us name $\mathbb{M}$ the resulting classes of equivalences. We define an order relation on $\mathbb{M}$ by:

$$\mu > v \Leftrightarrow \exists (M,N) \in \mu \times v / M \sim N(n_0) = \{n \in N / n < n_0\}.$$

Show that:

a) There is a minimal element in $\mathbb{M}$;

b) Any couple of elements in $\mathbb{M}$ can be compared;

c) $\mathbb{M}$ is well ordered;

d) $\mathbb{M}$ is non-enumerable;

e) Any non-enumerable set has a part that is equipotent to $\mathbb{M}$.

SOLUTION:

We first must check that what we have defined is an order relationship. The only point to check is that if X is a well-ordered set, then we cannot have $X \sim X(b)$: let us suppose that $X \sim X(b)$, let us call φ the strictly increasing bijection from X to $X(b)$, then we will have necessarily $\varphi(b) < b$ but if we consider the set $\{\varphi^n(b)\}_{n\in\mathbb{N}}$, we see that this set cannot have a minimum, which is impossible because X is a well-ordered set.

a) We consider $\mu_0 = \overline{\mathbb{N}}$.

If M is a well-ordered set, then it has a minimal element m_1. We can build a sequence m_n defined by:

$$m_{n+1} = \min M_n, M_n = M_{n-1}\setminus\{m_n\}\text{if } M_n \neq \emptyset \,, m_{n+1} = \emptyset\text{if } M_n = \emptyset \,, M_0 = M.$$

We put $m_\infty = \min(M\setminus\bigsqcup_{i=1,\dots,\infty}\{m_i\})$.

If $\{m_\infty\} = \emptyset$ then $M \sim \mathbb{N}$ and $M \in \mu_0$.

If $\{m_\infty\} \neq \emptyset$ then $M(m_\infty) = \{m \in M/m < m_\infty\}$ is equivalent to $\mathbb{N}$ through the bijection $n \to m_n$[28]. Therefore we have:

$$M(m_\infty) \in \mu_0 \implies \mu_0 \leq \overline{M}.$$

So that finally:

$$\mu_0 \text{ is the minimum element of } \mathbb{M}.$$

b) We will use transfinite induction (see Exercise 21.b).

We consider the set $S(L,M)$ whose elements are the pairs (A, φ_n) defined by:

$A \subset M, A = M'(m_0)$ where $M'(a) = M(a) \cup \{a\} = [M_{\min}, a]$ and φ is a strictly increasing bijection $A \to L(n)$, with the convention[29] that $n = \infty$ if φ is a bijection from A to L. Then we can partially order the pairs by making $(A, \varphi_n) \leq (B, \psi_m)$ if:
  - $A = L'(a_0)$,
  - $B = L'(b_0)$,
  - $A \subset B$,
  - $\psi_m|_B = \varphi_n$.

---

[28]This bijection is strictly monotonous because $(i < j) \implies (m_i < m_j)$

[29]This convention will be used only for the b)

Then $S(L,M)$ is inductively ordered because for any subset $\{A_\alpha, \varphi_n^\alpha\}_{\alpha \in U}$ of $S(L,M)$, we can define a function $\tilde{\varphi}$ from $\bigcup_{\alpha \in U} A_\alpha$ into L such that $\tilde{\varphi}|_{A_\alpha} = \varphi_n^\alpha$, and then $(\bigcup_{\alpha \in U} A_\alpha, \tilde{\varphi})$ is an upper bound for the subset.

We distinguish the following cases:

1) L has a maximum element but M does not;

2) Symmetric case of 1);

3) Both L and M have a maximum element;

4) neither L or M have a maximum element.

case 1) and 2).

Let m_m be a maximal element of M, m_m is in fact the maximum element[30] (because M is well-ordered). We let (A_0, φ_0) be a maximum element in $S(L,M)$ (following Zorn's lemma, this element does exist). If φ_0 maps bijectively A_0 to $L(l_0)$ then:

- either $l_0 = l_\infty$ and φ_0^{-1} is a monotonous bijection from L into $A_0 = M'(m_0)$ and this will imply that $L \sim M'(m_0)$. From this, either $m_0 = m_\infty = m_m$, $M'(m_0) = M$ and $L = M$, or we can define $m_0 + 1 \le m_m$ in M so that $L \sim M(m_0 + 1)$;

- either $l_0 < l_\infty$ then we must have in that case $A_0 = M$, because, if it is not the case, let us say that $A_0 = M'(m_0)$ we can define A_0' by $A_0' = A_0 \sqcup \{m_0 + 1\} = M'(m_0 + 1) = [M_{\min}, m_0 + 1]$ $(m_0 + 1 \le m_m)$ and $l_m > l_0$ in L. We may extend φ_0 in φ_0' by making:

$$\varphi_0'|_{A_0} = \varphi_0, \varphi_0'(m_0) = l_m \text{ so that } (A_0', \varphi_0') > (A_0, \varphi_0) \text{ - which is impossible.}$$

We see in all cases that we can compare L and M.

Case 3).

L has a max l_m and M has a max m_m. Let (A_0, φ_0) be the max element of $S(L,M)$ and let $\varphi(A_0) = L(l_0)$.

- If $A_0 \ne M$ and $l_0 = l_m = l_\infty$, then φ^{-1} realize the equivalence $L \sim M(m_0)$.

---

[30]We may refer to it as the max element of the set in what follows

- if $A_0 \neq M$ and $l_0 \neq l_m$, then if $A_0' = A_0 \sqcup \{m_0 + 1\}$ and φ_0' defined by: $\varphi_0'|_{A_0} = \varphi_0$ and $\varphi_0'(m_0) = l_0$, then we will have $(A_0', \varphi_0') > (A_0, \varphi_0)$, which is impossible.

Case 4).

If both L and M do not have a max element, then if (A_0, φ_0) is a max element in $S(L, M)$ and $\varphi(A_0) = L(l_0)$ then either $l_0 = l_\infty$, or $A_0 = M$. If not, we can define $l_0 + 1$ and $m_0 + 1$, again we define $A_0' = [M_{\min}, m_0 + 1]$ and the function $\tilde{\varphi}_0$ by: $\tilde{\varphi}_0|_{A_0} = \varphi_0$, $\tilde{\varphi}_0(m_0 + 1) = l_0$ and this leads to the fact that $\tilde{\varphi}_0$ establishes a correspondence f, A_0' to $L(l_0 + 1)$ so that $(A_0', \tilde{\varphi}_0) > (A_0, \varphi_0)$ which is impossible.

c) Let us show that $\mathbb{M}$ is well-ordered. We consider, as suggested in the hint, a set $\mathbb{M}_0 \subset \mathbb{M}$, if $\mu \in \mathbb{M}_0$, we choose a representant M of μ in $\mathbb{M}_0$ (e.g. $\overline{M} = \mu$). Let $\mu_1 \in \mathbb{M}_0$, $\overline{M_1} = \mu_1$, then either $\mu_1 \geq \mu$ for all μ_1 and then μ is the minimum of $\mathbb{M}_0$, or we can find $\mu_1 \in [\mu_0, \mu)$ - that is to say, we can find m_1 in M such that $M_1 \sim M(m_1)$. We then can build a function F from $\mathbb{M}_0 \cap [\mu_0, \mu)$ into M by making $F(\mu_1) = m_1$.

We consider $\mu_2 \in \mathbb{M}_0 \cap [\mu_0, \mu)$ such that $\mu_1 < \mu_2$ in $\mathbb{M}_0 \cap [\mu_0, \mu)$, if $\overline{M_1} = \mu_1$ and $\overline{M_2} = \mu_2$ and $F(\mu_1) = m_1$, $F(\mu_2) = m_2$, we want then to prove that $m_1 < m_2$.

We know that $M_1 \sim M_2(n_2)$ and we have the following situation (Fig 2) :

φ_1 realize the equivalence $M_1 \sim M(m_1)$, φ_2 realize the equivalence $M_2 \sim M(m_2)$, φ_{12} realize the equivalence $M_1 \sim M_2(n_2)$ where φ_1, φ_2 and φ_{12} are strictly increasing bijections. It is clear that $\varphi_2(n_2) < m_2$ since $M_2 \neq M_2(n_2)$. But $M_1 \sim M(\varphi_2(n_2))$ by $\varphi_2 \circ \varphi_{12}$ so that $M(\varphi_2(n_2)) \sim M(m_1)$.

We must then have $\varphi_2(n_2) = m_1$ otherwise, if we define φ as $\varphi = \varphi_2 \circ \varphi_{12} \circ \varphi_1^{-1}$ then we can find x such that $\varphi(x) < x$ (the value of x depending if $m_1 < \varphi_2(n_2)$ or $m_1 > \varphi_2(n_2)$) and then the subset of M defined by $\{\varphi^n(x), n \in \mathbb{N}\}$ will have no minimum element which is impossible since M is well-ordered.

So that we must have $m_1 = \varphi_2(n_1) < m_2$ and F is a strictly increasing mapping. We know that M is well-ordered, then the image $F(\mathbb{M}_0 \cap [\mu_0, \mu))$ must have a minimum, let us say m_0, but then $F^{-1}(m_0)$ is the minimum of

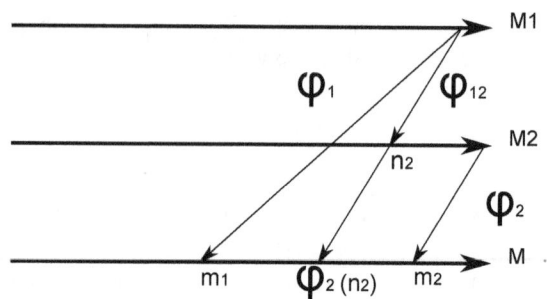

Figure 2: Demonstration in 15c)

$\mathbb{M}_0 \cap [\mu_0, \mu)$ because F is a strictly increasing mapping. This shows that $\mathbb{M}$ is well-ordered since any subset $\mathbb{M}_0$ have a minimum element.

d) Let us show that $\mathbb{M}$ is a non enumerable set. If we suppose that $\mathbb{M}$ is enumerable then we can order its element by $\mathbb{M} = \{\mu_i\}_{i=1}^{i=\infty}$. If we choose the sets $M_i, i = 1, \ldots, \infty$ such that $\overline{M_i} = \mu_i$ then we can form the set M by: $M = \bigcup_{i=1}^{i=\infty} M_i$. As suggested in the hint we define in M the relation $x \leftrightarrow y$ by:

$\quad x \leftrightarrow y \Leftrightarrow \exists f_{ij}, M_i, M_j, n_{ij}$ such that: $x \in M_i, y \in M_j(n_{ij})$ and f_{ij} is a strictly increasing bijection $M_i \sim M_j(n_{ij})$ or $M_j \sim M_i(n_{ij})$.

From a graphical point of view, the quotient space is obtained by "stacking" the sets M_i's such that $i < j \implies \overline{M_i} < \overline{M_j}$ and by taking the "steps" as classes.

Let us show that $x_i \leftrightarrow x_j$ is an equivalence relationship:

1) $x_i \leftrightarrow x_i$, because, trivially $M_i = M_i$,

2) $x_i \leftrightarrow x_j \implies x_j \leftrightarrow x_i$ by the (symetric) definition of $\leftrightarrow$,

3) if we have $x_i \leftrightarrow x_j$ and $x_j \leftrightarrow x_k$, then, if $x_i \in M_i$, $x_j \in M_j(n_{ij}), x_k \in M_k(n_{jk}), M_i \sim M_j(n_{ij}), M_j \sim M_k(n_{jk})$.

We then have $x_j < n_{ij}$ and $n_{ik} = \varphi_{jk}(n_{ij})$ (see part c)).

So that $x_k \in M(n_{ik})$ and $x_i \leftrightarrow x_k$ (in fact $\varphi_{ik} = \varphi_{jk} \circ \varphi_{ij}$ thus we could define the sets M_i's in term of a category with the morphism being the φ_{ij} but that is useless here).

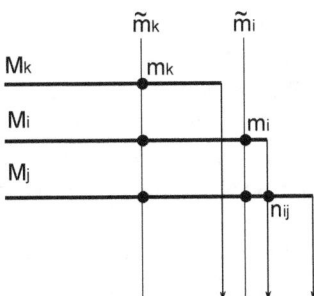

Figure 3: Equivalence classes in $\tilde{X}$

The reasoning in the other cases is similar so that $x_i \leftrightarrow x_j$ is an equivalence relationship in M. We let $\tilde{X}$ be the corresponding quotient space. $\tilde{X}$ is totally ordered, for if $(\tilde{m}_1, \tilde{m}_2) \in \tilde{X}$ we can define $\tilde{m}_1 < \tilde{m}_2$ by $\varphi_{12}(m_1) < m_2$ over M_2. This relation is independant of the choice of the representants $m'_1 \in M'_1, m'_2 \in M'_2$. If we have two other representants $\tilde{m}'_1$ and $\tilde{m}'_2$ such that $(\tilde{m}'_1 = (\tilde{m}_1$ and $\tilde{m}'_2) = \tilde{m}_2)$ then we have the following situation shown in Fig 4.

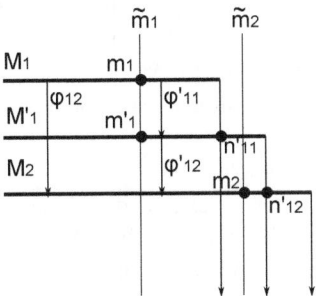

Figure 4: Definition of an order over $\tilde{X}$

We can find φ'_{11}, a strictly increasing mapping: $M_1 \sim M'_1(n'_{11})$ and φ'_{12}, a strictly increasing mapping: $M'_1 \sim M_2(n'_{12})$. (or we may permute M_1 and M'_1 and/or M_2 and M'_2). We know that $\varphi_{12} = \varphi'_{12} \circ \varphi'_{11}$ and we have: $m'_1 = \varphi'_{11}(m_1)$ and $\varphi_{12}(m_1) = \varphi'_{12}(m'_1)$ so that $\varphi'_{12}(m'_1) < m_2 \Leftrightarrow \varphi_{12}(m_1) < m_2$.

We can do the same reasoning for m_2 and m'_2 and this shows that the choice of the representant does not impact on the definition of our order over $\tilde{X}$.

So that M is provided with a total order induced by the total order of each of the M_i's (see Exercise 21.a), construction of $X = \bigcup_{\alpha \in A} X_\alpha$).

Let us now consider $\tilde{S}$, a part of $\tilde{X}$. If x_i is the minimum for the set $S \cap M_i = S_i$, then (same reasoning as in Exercise 21 a)), $\tilde{x}_i$ is the minimum for $\tilde{S}$ because $\tilde{x}_i$ does not depend of the index i.

This shows that $\tilde{X}$ is well-ordered by the order we have defined.

We note that $\tilde{X}$ is non-finite (enumerable meaning here coutable non-finite) since it contains the $\{\tilde{n}_{ij}\}$'s $j = 1, \dots, n, \dots \infty$ and this sequence is is an infinite sequence.

We conclude that we have $\overline{\overline{\tilde{X}}} \in \mathbb{M}$, $\tilde{X}$ being enumerable since it is an enumerable union of enumerable sets (See Exercise 23).

But if we consider the mapping $M_i \xrightarrow{f_i} \bigcup_{n=1}^{\infty} M_i \xrightarrow{\tilde{g}} \tilde{X}$, where f_i is the injective canonical mapping $M_i \to \bigcup_{i=1}^{i=\infty} M_i$ and $\tilde{g}$ the surjective canonical mapping $M \to \tilde{X}$, then $\tilde{g}_i$ is an injective mapping because $x, y \in M_i, x \neq y$ will imply that $\tilde{g}_i(x) \neq \tilde{g}_i(y)$ in $\tilde{X}$, it is strictly increasing and furthermore $\tilde{g}_i(M_i) = \tilde{X}(n_{ij})$, where n_{ij} is defined by $M_i \sim M_j(n_{ij})$ (and this is independent of j), we then have anyway that for all i, $\overline{\overline{M_i}} < \tilde{X}$ or: $\forall i \geq 1, \mu_i < \overline{\overline{\tilde{X}}}$ which is impossible because $\overline{\overline{\tilde{X}}} = \mu_j$ for an index j and this would mean that $\mu_j < \mu_j$.

Then $\mathbb{M}$ is non-denumerable.

e) We want to show that $\mathbb{M} = \omega^1$, ω^1 being *the first uncountable ordinal* (e.g. the first non-enumurable ordinal). First we need to define the class Ω of ordinals [31] as the class of well-ordered sets $(X, \leq)$ modulo a bijective

---

[31]Ω is a class and not a set but as we restrict ourselves to an intial segment of Ω, this does not matter here, however we can replace the term "class of ordinals" by "set of ordinals" without changing anything to the reasoning for $\mathbb{M}$

increasing mapping. We see that in fact, $\mathbb{M}$ is a part of Ω, that is to say the restriction of Ω to enumerable non-finite sets.

Ω can be equivalently represented as follows (Von Neumann approach): first there is the emptyset $\emptyset$, then $\{\emptyset, 0\}$, then $\{\emptyset, 0, 1\}$, ...E_N... where $E_N = \{\emptyset, 0, 1, \ldots, N\}$, after the sets E_N will come the set ω^0, that is to say $\mathbb{N}$ provided with the natural order, then $\omega^0 + 1, \ldots \omega_0 + N, \ldots M \times (\omega^0)^P + N$, etc... We will meet in Ω, the first uncountable ordinal at $(\omega^0)^{\omega^0}$. So that the elements before ω^1 in Ω can be described by:

$$\emptyset, \{\emptyset, 0, 1\}, \ldots, \{\emptyset, 0, 1, \ldots, N\}, \ldots, \omega^0, \ldots, (M \times (\omega^0)^P + N)_{M, N, P \in \mathbb{N}},$$

.... Ω will contain after ω^1 the elements $\omega^{0^{\omega^{0 \ldots \omega^0}}}$, etc... untill ε_0 is reached ("omega nought"), etc...

The Von Neumann approach consists in picking sets that will be representant of equivalence classes. One extremely interesting property of Ω is the following:

$$\forall \omega \in \Omega, [0, \omega) = \omega.$$

Indeed, we see that $[\emptyset, \{\emptyset, 0\}) = \emptyset, [\emptyset, \{\emptyset, 0, 1\} = \{\emptyset, 0\}$ and that more generally: $[\emptyset, E_N) = E_N$ for all $N \geq 0$ so that, finally, $[\emptyset, \omega^0)$ is equal to ω^0.

Let us prove this result by transfinite induction (See Exercise 21.b). We note E the subset of Ω where this property holds.

First we note that we have $[\emptyset, \omega) + \{\omega\} = [\emptyset, \omega]$.

We see that $[\emptyset, \{\emptyset, 0\}) = \emptyset$ as noted before so that $\emptyset \in E$.

If $\forall \omega < \omega'$, $\omega \in E$, then $[\emptyset, \omega] = [\emptyset, \omega) + \{\omega\} = \omega + 1$ [32], $\forall \omega < \omega'$, we have that $[\emptyset, \omega') = \bigcup_{\omega < \omega'} [\emptyset, \omega]$ so that:$[\emptyset, \omega') = \bigcup_{\omega < \omega'} \omega + 1$.

We consider $x \in \Omega$ and $\mathscr{P}(\omega)$, the following property: $\omega < \omega' \Rightarrow x > \omega$. If $\mathscr{P}(\omega)$ holds for every $\omega \in \Omega$, then we must have $x \geq \omega'$: indeed, if $X_{\omega'} = \Omega - [\emptyset, \omega')$ then $\omega' = \min(X_{\omega'})$ and if $x \neq X_{\omega'}$ then $x < \omega'$ but, since we have also $\mathscr{P}(x)$, this implies that $x < x$, which is impossible.

Then we put $x = \bigcup_{\omega < \omega'} \omega + 1$, we see easily that $x > \omega, \forall \omega < \omega'$ then we have as mentioned before, $x \geq \omega'$.

---

[32] Since Ω is itself a well-ordered set, every element $\omega \in \Omega$ has a unique successor $\omega + 1$ defined as $\min\{\Omega - [\{\emptyset\}, \omega]\}$

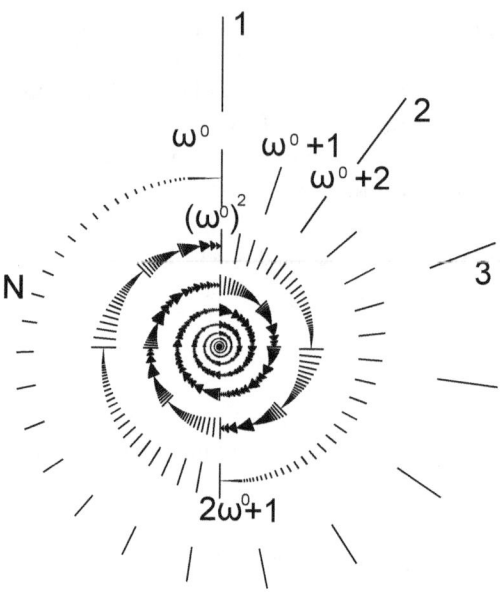

Figure 5: Graphical representation of the ordinals up to ω^1, $\mathbb{M}$ is the part starting from ω^0 and going into the "eye" of the spiral

On the other hand, we have that $\omega < \omega' \implies \omega + 1 \leq \omega'$ so that $x = \bigcup_{\omega < \omega'} \omega + 1 \leq \omega'$.

Finally we have $x = \omega'$ what proves that $[\emptyset, \omega') = \omega'$ and $\omega' \in E$. Then by transfinite induction: $E = \Omega$.

An intuitive vision of the property $[\emptyset, \omega) = \omega$ is that there are, in fact, exactly "ω" ordinals who are inferior to ω.

Now, we return to our goal which is to show that $\mathbb{M} = \omega^1$ (at least from a cardinal point of view): the set of non-enumerable ordinals (let us say Ω') has a minimum for Ω is well-ordered by the order defined in $\mathbb{M}$ and extended to Ω (the proof that this is a well-order is similar to the demonstration used for $\mathbb{M}$, the well-ordering of the elements of Ω will define a well-ordering in Ω). We note $\omega^1 = \min(\Omega')$. We then must have $\omega^1 \leq \mathbb{M} + \omega^0$.

Idea 1): As mentioned before, we just note that $\mathbb{M}$ is the set of enumerable ordinals that is to say $\mathbb{M} = [\omega^0, \omega^1)$. That imply that $\mathbb{M} = -\omega^0 + \omega^1$. This is enough to show that all non-enumerable set X will be such that $card(X) \geq card(\mathbb{M})$.

Idea 2):

We wish to prove that $\forall \mu \in \mathbb{M}, -\omega^0 + \mu = (\mu)$ [33] by transfinite induction (see Exercise 21 b).

1) This is true for $\mu = \omega^0$ since $\mathbb{M}(\omega^0) = \emptyset = -\omega^0 + \omega^0$.

2) If this true for all $\mu < \mu_1, (\mu, \mu_1) \in \mathbb{M}$, then $\mathbb{M}'(\mu)$ [34] will be such that:

$$\mathbb{M}'(\mu) = \mathbb{M}(\mu) + \{\mu\} = -\omega^0 + (\mu + 1).$$

With the same arguments than for the demonstration that $[0, \omega) = \omega$ in Ω [35]. We have:

$$
\begin{aligned}
\mathbb{M}(\mu') &= \bigcup_{\omega^0 \leq \mu \leq \mu'} \mathbb{M}'(\mu), \\
&= \bigcup_{\omega^0 \leq \mu \leq \mu'} [\omega^0, \mu), \\
&= -\omega^0 + \bigcup_{\omega^0 \leq \mu \leq \mu'} (\mu + 1).
\end{aligned}
$$

Thus the element:

$$-\omega^0 + \bigcup_{\omega^0 \leq \mu \leq \mu'} (\mu + 1)$$

will be equal to: $x = -\omega^0 + \mu^1$.

---

[33] We note $-\omega^0 + \mu$ the set μ with its first ω^0 element being removed. We also note $\omega^0 + \mu$ the set μ with ω^0 new elements added to the beginning.

[34] This set being defined as $[\omega^0, \mu]$ in $\mathbb{M}$

[35] Again, $[\omega^0, \mu]$ is being defined in $\mathbb{M}$, not in Ω

This proves the result by transfinite induction.

Now, we know that $\omega' \leq \mathbb{M}$. Let us suppose that $\mathbb{M} \neq \omega'$ then $\omega' \in \mathbb{M}$ since $\omega' > \omega^0$, so that $-\omega^0 + \omega' = \mathbb{M}(\omega')$ [36] or equivalently $\omega' = \omega^0 + \mathbb{M}(\omega')$. But we know that $\mathbb{M}(\omega')$ is enumerable so that this implies that ω' is also enumerable, which is enumerable. We must have, therefore, $\mathbb{M} = \omega'$.

---

[36]In fact, it is easy to check that $-\omega^0 + \omega' = \omega'$ since ω' is the least uncountable ordinal

16/828 ¶

Subject: Properties of $\omega^1 \times [0,1[$

Let $\mathbb{M} \approx \omega^1$ be the totally ordered set defined in the exercise 15 (e.g. the first uncountable ordinal). We define $\mathbb{U} = \mathbb{M} \times [0,1[$ and we provide $\mathbb{U}$ with a lexicographic order : if $a = (\mu, x), b = (\nu, y)$, then $a > b$ means that either $\mu > \nu$ or $\mu = \nu$ and $x > y$.

Show that any initial interval (see Exercise 15) from the set $\mathbb{U}$ is equivalent to the interval $[0,1[$ while the whole set $\mathbb{U}$ is not.

SOLUTION:

We can represent the set $\mathbb{U}$ as the non-enumerable ω^1-"gluing" of copies of $[0,1[$, the points $(\mu, 1)$[37] and $(\mu + 1, 0)$ being the points "glued" together[38]:

$$\omega^0 \leftrightarrow [0,1[\leftrightarrow \omega^0 + 1 \leftrightarrow \ldots - [0,1[\leftrightarrow \omega^0 + n \leftrightarrow [0,1[\leftrightarrow \ldots \leftrightarrow \mu \leftrightarrow [0,1[\leftrightarrow \mu + 1 \leftrightarrow [0,1[\leftrightarrow \ldots \omega^1.$$

Let $\mathbb{U}(\mu, x) = [(\omega^0, 0), (\mu, x))$ be an initial interval of (μ, x) then $\mathbb{U}(\mu, x)$ is equivalent to:

$$\mathbb{U}(\mu, x) \sim (\bigsqcup_{\omega^0 \leq \nu < \mu} [0,1[) \sqcup [0,x[$$

or equivalently from the properties of Ω (see Exercise 15 e)):

---

[37] $(\mu, 1)$ is not an element of $\mathbb{U}$ but we can identify it to the element $(\mu + 1, 0)$
[38] The symbol $\leftrightarrow$ represents this gluing on the diagram

$$\mathbb{U}(\mu,x) \sim ((-\omega^0 + \mu) \times [0,1[) \sqcup [0,x[.$$

1)If $-\omega^0 + \mu$ is finite, then: $-\omega^0 + \mu = \{\emptyset, 0, 1, \ldots, N\}$ and $\mathbb{U}(\mu,x)$ will consist in the gluing of N copies of $[0,1[$, itself glued with $[0,x[$ so that $\mathbb{U}(\mu,x) \sim [0, N+x[$ (the bijection f being defined by $f(\mu_i, [0,1[) = [i, i+1[$ (We also name $(\mu_i, [0,1[) = [0,1[_i$ the ith copy of $[0,1[$ being glued) and $f((\mu, [0,x[)) = [N, N+x[$ (with the convention that $[N, N+x[= \emptyset$ if $x = 0$) then by using the mapping $u \to \frac{u}{N+x}$ we see that $\mathbb{U}(\mu,x) \sim [0,1[$.

2)If $-\omega^0 + \mu$ is enumerable (non-finite) then:

$$-\omega^0 + \mu = \{\emptyset, 0, 1, \ldots, N, \ldots\}.$$

Let us note $\{x_1, \ldots, x_N, \ldots\}$ the elements of $\mu - \omega^0$ well-ordered e.g. such that $x_i < x_j$ for $i < j$. We can add $[0,x[\sim [0,1[$ to the set by adding formally a point $x_0 < x_1$ to it. Then we see that $\mathbb{U}(\mu,x) \sim \mathbb{R}^+$ indeed we can define a bijection f by: $f((x_i, [0,1[)) = [i, i+1[, \forall i \geq 0$. We know that $[0,1[\sim \mathbb{R}^+$ (by taking $u \to \frac{u}{1-u}$ or $u \to \tan \frac{2u}{\pi}$ for example) so that, again $\mathbb{U}(\mu,x) \sim [0,1[$.

But the whole set $\mathbb{U}$ is not equivalent (as an ordered set) to an interval of $\mathbb{R}$. First, we note that the cardinality of $\omega^1 - \omega^0$, and hence the cardinality of ω^1 is the cardinality of $\mathbb{R}$ (that is to say Aleph-1, $\aleph^1$).

Indeed, from our construction of Ω, the limit of the sets $M \times (\omega^0)^P + N$ will be $(\omega^0)^{\omega^0}$ provided with a lexicographical order (no non-enumerable ordinal can be $< (\omega^0)^{\omega^0}$) furthermore this set is equipotent to $[0,1]$.

We could, at first, try to argue that if we could find a bijective, strictly increasing mapping φ from ω^1 to $[0,1[$ (in this context $[0,1[$ is provided with its natural order), then the equivalence $\mathbb{U} \sim [0,1[$ would imply that we could find a bijective strictly increasing mapping from $[0,1]$ to $[0,1[^2$: e.g. from $\varphi : \omega^1 \to [0,1[$ we could define $\tilde{\varphi} : \mathbb{U} \to [0,1[^2$ by making $\tilde{\varphi}(\mu,x) = (\varphi(\mu),x)$. Then if we can find ψ strictly increasing such that: $\psi : \mathbb{U} \to [0,1[$ then the mapping $\tilde{\varphi} \circ \psi^{-1}$ will show that $[0,1] \sim [0,1[^2$. This mapping would be right and left continuous and it is known that such mapping must be totally discontinuous.

But we cannot find φ such that $\omega^1 \sim [0,1[$ because otherwise $[0,1[$ would be well-ordered, which is not the case. Instead of this we can find strictly increasing bijections such that $\omega^1 \sim \omega^{0\omega^0}$ (provided with the lexicographic order) and again $(\omega^0)^{\omega^0} \sim 2^{\omega^0}$, 2^{ω^0} being also provided with the lexicographic order. 2^{ω^0} is equivalent as an ordered set to C, the Cantor Set (see Exercise 42) and, in general to any all compact, perfect and totally disconnected subsets of $\mathbb{R}$ (see again Exercise 42) but this is as far as we could go, so our idea (which was to try to show that $\mathbb{U} \sim [0, 1[$ imply the existence of a left and right continous mapping $[0, 1[\rightarrow [0, 1[^2)$will not work in that context. On the meantime we have seen that $\mathbb{U} \sim \omega^{0(\omega^0)} \times [0, 1[\sim 2^{\omega^0} \times [0, 1[\sim \mathfrak{C}_1 \times [0, 1[$.

In fact if we look a bit closer, we can simply use the fact that a strictly increasing bijection from $\omega^1 \times [0, 1[$ to $[0, 1[$ will map intervals of $[0, 1[$ into intervals of $[0, 1[$. Indeed, let us suppose that we can find a correspondence $f : \omega^1 \times [0, 1[\sim [0, 1[$ then we will define f_ω by $f_\omega(x) = f(\omega, x)$. We have that $f_\omega([0, 1[)$ is the interval $[f_\omega(0), f_\omega(1)[$, indeed, if $u \in [f_\omega(0), f_\omega(1)[= [f(\omega, 0), f(\omega, 1)[$ and $u = f(\mu, x')$ then we have $f(\omega, 0) \leq f(\mu, x') < f(\omega, 1)$ and from this we deduce that $\mu = \omega$ and $x' \in [0, 1[$, that is to say $(\mu, x') \in (\omega, [0, 1[)$. On the other hand, we see that obviously $\forall x \in [0, 1[$, $f_\omega(0) \leq f_\omega(x) < f_\omega(1)$, so that $f_\omega([0, 1[) = [f_\omega(0), f_\omega(1)[$.

Then we will have:

$$[0, 1[= \bigsqcup_{\omega < \omega^1} [f_\omega(0), f_\omega(1)[.$$

If we let $\delta_\omega = f_\omega(1) - f_\omega(0)$, then $\delta_\omega > 0, \forall \omega$ and for all $\varepsilon > 0$, the set X_ε defined by $X_\varepsilon = \{\omega < \omega^1, \delta_\omega \geq \varepsilon\}$ will be finite ($card(X_\varepsilon) \leq [1 + 1/\varepsilon])$ but we have: $\omega^1 = \bigcup_{n=1}^{n=\infty} X(\frac{1}{n})$ and this is impossible since this would mean that ω^1 is enumerable or finite.

17/828 ¶

Subject: Topological properties of the Alexandrov Line $\mathbb{U}_0$ (Open Long Ray)

Let $a_0 = (\mu_0, x_0)$ the minimum point of the set $\mathbb{U}$ from Exercise 16. We define a topology over $\mathbb{U}_0 = \mathbb{U} \setminus \{a_0\}$ by considering as base of open neighborhoods the "intervals" $(a,b) = \{c \in \mathbb{U}_0 : a \leq c \leq b, c \neq a, c \neq b\}$ (The Order Topology). Show that:

a) every point $a \in \mathbb{U}_0$ have a neighborhood that is homeomorphic to an ordinary interval;

b) the topological space $\mathbb{U}_0$ is connected and is not homeomorphic to an ordinary interval.

The space $\mathbb{U}_0$ is named the *Alexandrov Line* (or Open Long Ray) and it is an example of a 1-dimensional manifold that do not have any enumerable base of open neighborhood[39].

SOLUTION:

a) We see that $\mathbb{U}_0 = \mathbb{U} - \{(\mu_0, 0)\} = \mathbb{U} - \{(\omega^0, 0)\}$.

i) It is rather straightforward in the case where $x \neq 0$, to see that we can find a neighborhood $V(\mu, x)$ of (μ, x) such that $V(\mu, x)$ is homeomorphic

---

[39]In the context of the exercise, the Alexandrov line $\mathbb{U}_0$ is also generally known as The Open Long Ray. $\mathbb{U}$ is the closed long ray. We can create an other entity known as the Long Line, in order topology by gluing together a reversed copy of the Open Long Ray and a copy of the Closed Long Ray

to $\mathbb{R}$, indeed,we take $\varepsilon > 0$ such that both $x - \varepsilon$ and $x + \varepsilon$ lie in $]0, 1[$ and we set $V_\varepsilon(\mu, x)$ to be equal to:

$$V_\varepsilon(\mu, x) = \{(\mu, x'), x - \varepsilon < x' < x + \varepsilon\} = ((\mu, x - \varepsilon), (\mu, x + \varepsilon)),$$

then $V_\varepsilon(\mu, x)$ is homeomorphic to $]x - \varepsilon, x + \varepsilon[$ (and thus to $]0, 1[$).

ii) If x=0, then $V_\varepsilon(\mu, 0) = ((\mu_0, \varepsilon), (\mu, \varepsilon))$ is a neighborhood of of $(\mu, 0)$. If $-\mu_0 + \mu$ $(\mu_0 = \omega^0)$ is finite with cardinality = N then $V_\varepsilon(\mu, 0)$ is homeomorphic to $[0, N + 1[$ (see Exercise 16) and if $\mu - \mu_0$ is enumerable, non-finite, then we put $\mu - \mu_0 = \{x_i, i \geq 0\}$ $(x_i < x_j$ if $i < j)$ and we define, similarly to Exercise 16, the homeomorphisms, φ, by: $\varphi(\mu_0,]\varepsilon, 1[) =]0, 1[$, $\varphi(x_i, [0, 1[) = [i, i+1[$ for $i > 0$ and $\varphi(\mu, [0, \varepsilon[) = [1, 1 + \varepsilon[$. Then we have and homeomorphisms from $V_\varepsilon(\mu, 0)$ to $]0, 1[\sqcup[1, 1 + \varepsilon[$ that is to say that $V_\varepsilon(\mu, 0)$ is homeomorphic to $]0, 1[$.

This prove that $\mathbb{U}_0$ is a 1-dimensional manifold.

b) Let us show that $\mathbb{U}_0$ is connected. In fact it is arc-connected. If (μ_1, x_1) and (μ_2, x_2) are two elements of $\mathbb{U}_0$ then we can easily define a path from (μ_1, x_1) to $(\mu_1, 0)$ and from (μ_2, x_2) to $(\mu_2, 0)$.

Then we can connect $(\mu_1, 0)$ and $(\mu_2, 0)$.

We define the paths $f_i : [0, 1] \to [(\mu + i, 0), (\mu + i + 1), 0]$ for $i \geq 0$, by: $f_i(t) = (\mu + i, t)$.

i) If $-\mu_1 + \mu_2 = \{x_1, \dots, x_N\}$, then we may define $\tilde{f}$ by:

$$\tilde{f}|_{[1-\frac{1}{n}, 1-\frac{1}{n+1}]}(t) = f_{n-1}((n^2 + n)t - (n^2 - 1)), \text{ for } n = 1, \dots, N.$$

Then $f(t) = \tilde{f}(\frac{N}{N+1}t)$ is a continuous function $[0, 1] \to \mathbb{U}_0$ such that $f(0) = (\mu_1, 0)$ and $f(1) = (\mu_2, 0)$, therefore it is a path between the two points.

ii) If $-\mu_1 + \mu_2$ is enumerable non-finite, then we can (again as in Exercise 15) define $\tilde{f}$ from $[0, 1]$ to $\mathbb{U}_0$ by:

$$\tilde{f}|_{[1-\frac{1}{n}, 1-\frac{1}{n+1}]} = f_{n-1}((n^2 + n)t - (n^2 - 1))$$

and

$$\tilde{f}(1) = (\mu_2, 0).$$

Again, $\tilde{f}$ will be continuous at $t = 1$: indeed for all neighborhoods $V_\varepsilon(\mu_2, 0) = ((\mu', \varepsilon), (\mu'', \varepsilon))$, $\mu' \le \mu_2 \le \mu''$) we can find $\delta > 0$ such that $\tilde{f}(]1 - \delta, 1[) \subset V_\varepsilon(\mu_2, 0)$.

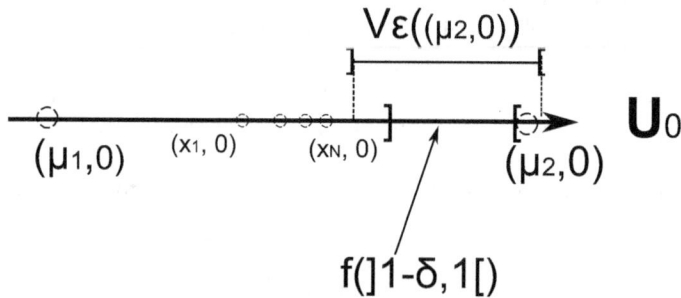

Figure 6: Continuity of $\tilde{f}$ in t=1

Let us now show that $\mathbb{U}_0$ is not homeomorphic to $\mathbb{R}$ (or $]0, 1[$). If this would be the case, then $\mathbb{U}_0$ would have an enumerable base of neighborhood (from the base of the topology in $\mathbb{R}$) but on the same way the set ω^1 is totally disconnected because of its well-order so that all base of neighborhood of $\mathbb{U}_0$ must have a cardinality at least the cardinality of ω^1. This means that all base of neightborhood in $\mathbb{U}_0$ are non enumerable, but if B if the set $\{]\frac{1}{n}, \frac{1}{n+1}[, n \ge 1\}$, and F the homeomorphism from $\mathbb{U}_0$ to $\mathbb{R}$, then $F^{-1}(B)$ will be an enumerable base of neightborhood of $\mathbb{U}_0$, what we have shown to be impossible.

$\mathbb{U}_0$ is a 1-dimensional manifold named the Alexandrov Line (or the Open Long Ray) that do not have an enumerable base of neightborhood (e.g. is not second countable) and thus it is not metrizable from Urysohn's metrizability Theorem . We also note that, with the use of the same theorem, this also imply that $\mathbb{U}_0$ cannot be homeomorphic to any subspace of the Hilbert Cube $\prod_{i=1}^{i=\infty}[0, 1]$.

18/828 ℬ

Subject: Maximum and maximal elements in a square

Show that the set of disks included in a given square of $\mathbb{R}^2$ have a maximal element but do not have a maximum element for inclusion.

SOLUTION:

First solution:

Let $\Gamma(S)$ be the set of (closed) disks included in a given square $\mathbf{S}$, then the subset T(S) of $\Gamma(S)$ containing the disks who are tangent to at least 2 sides of $\mathbf{S}$ is the set of maximal elements of $\Gamma(S)$. Indeed , if $D_m \in T(S)$ is such a disk, then we can suppose that D_m have tangency contact points with $\mathbf{S}$ in $\{A, B\}$.

If $D \in \Gamma(S)$ is such that $D_m \subset D$ then $\{A, B\} \subset D$. It is "geometrically obvious" that this implies that $\{A, B\}$ are tangency contact points between D and $\mathbf{S}$ (see Fig. 8).

But knowing two different lines and the two tangency contact points, we can build a only unique circle so that $D = D_m = D_{\max}$ and so that $D_{\max}$ is a maximal element of $\Gamma(S)$.

Conversely, we can build an application $\Gamma(S) \to T(S)$ such that $D \subseteq \varphi(D)$ (see Fig. 9) and $D = \varphi(D)$ if an only if $D \in T(S)$.

So that T(S) describes the set of maximal elements.

If D_m is a maximum element, then $D_m \subseteq \varphi(D_m)$, so that we must have $\varphi(D_m) = D_m$ or equivalently $D_m \in T(S)$. From "obvious geometric" considerations, we see that, given a disk D of $T(S)$ with contact points (A,B) we can always find an other circle D' in T(S) with contact points (A',B') so that $D \cap D' = \emptyset$ (see Fig. 8).This implies that there is no maximum element in $\Gamma(S)$.

Second solution:

If we want to give a somewhat less "descriptive" proof of this, then we might proceed the following way: if (x,y,r) defines a circle of center (x,y) and radius r in the coordinate system made by the two non-diagonal symmetric axis of the square S. We suppose that S have sides of length 1. $\Gamma(S)$ is the set of circles (x,y,r) satisfying:

$$|x| \leq \frac{1}{2}, |y| \leq \frac{1}{2},$$
$$r \leq \frac{1}{2} - \max(|x|,|y|),$$

and T(S) is defined by circles such that:

$$|x| = |y|, r = \frac{1}{2} - |x|, |x| \leq \frac{1}{2}.$$

For a given circle (x,y,r) of $\Gamma(S)$, we define a map φ from $\Gamma(S)$ to T(S) such that $\varphi(x,y,r) = (X,Y,R)$ where X,Y and R are computed as follows[40]:

$$
\begin{aligned}
X &= sgn(x)(\min(|x|,|y|)), \\
Y &= sgn(y)(\min(|x|,|y|)), \\
R &= \frac{1}{2} - \min(|x|,|y|).
\end{aligned}
$$

If (u,v) is a point inside (x,y,r) then $d((X,Y),(u,v)) \leq d((X,Y),(x,y)) + d((x,y),(u,v))$.

---

[40] $sgn(x) = 1$ if $x > 0$, $sgn(0) = 0$, $sgn(x) = -1$ if $x < 0$

By definition:

$$d((x,y),(u,v)) \le r \le \frac{1}{2} - \max(|x|,|y|).$$

We also see that:

$$
\begin{aligned}
d((X,Y),(x,y))^2 &= (x - sgn(x)(\min(|x|,|y|)))^2 + (y - sgn(y)(\min(|x|,|y|)))^2, \\
&= (|x| - \min(|x|,|y|))^2 + (|y| - \min(|x|,|y|))^2 (\text{since } x = sgn(x)|x|), \\
&= (|x| - |y|)^2.
\end{aligned}
$$

So that:

$$d((X,Y),(x,y)) = ||x| - |y|| = max(|x|,|y|) - \min(|x|,|y|).$$

And finally:

$$
\begin{aligned}
d((X,Y),(u,v)) &\le \frac{1}{2} - \max(|x|,|y|) + \max(|x|,|y|) - \min(|x|,|y|), \\
&\le \frac{1}{2} - \min(|x|,|y|), \\
&\le R.
\end{aligned}
$$

This shows that $(x,y,r) \subseteq (X,Y,R)$.
Furthermore, we see that:

$$\varphi(D) = D, \quad (\forall D \in T(S))$$

and

$$D \subset \varphi(D), \quad (\forall D \in \Gamma(S) - T(S))$$

so that T(S) *contains* the set of maximal elements of $\Gamma(S)$. Besides, if there is a maximum element D_{mx} in $\Gamma(S)$, it must lie in $T(S)$ otherwise we would have: $D_{mx} \subset \varphi(D_{mx})$.

Let us show that $T(S)$ *is* the set of maximal elements of $\Gamma(S)$ by proving that there is no order relationship between any two elements of $\Gamma(S)$.

If $D_m(x_m, y_m, r_m) \in T(S)$ and $D(x, y, r) \in T(S)$ such that $D_m \subseteq D$. Then we must have $r \geq r_m$ and since $r = \frac{1}{2} - |x|, r_m = \frac{1}{2} - |x_m|, |x_m| \geq |x|$ (and $|y_m| \geq |y|$).

Since the length of the segment from the center of D to the center of D_m is $\sqrt{2}|x_m - x|$ and $D_m \subseteq D$ we have:

$$r_m + \sqrt{(2)}|x_m - x| \leq r.$$

That is to say:

$$\frac{1}{2} - |x_m| + \sqrt{(2)}(|x_m| - |x|) \leq \frac{1}{2} - |x|,$$

$$(\sqrt{(2)} - 1)(|x_m| - |x|) \leq 0.$$

The only way this can happen is that $x = x_m$, $y = y_m$, $r = r_m$ so that $D = D_m$. So that there is no order relation between any two distinct elements of $T(S)$: all elements of $T(S)$ are maximal. We know that a maximum element must be inside $T(S)$ so that $\Gamma(S)$ does not have a maximum element.

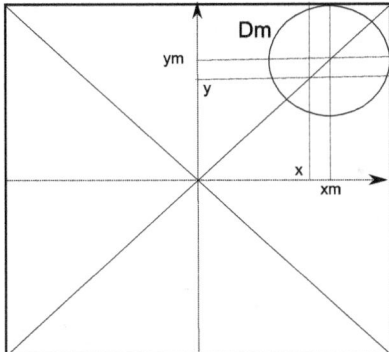

Figure 7: Analytic description of T(S).

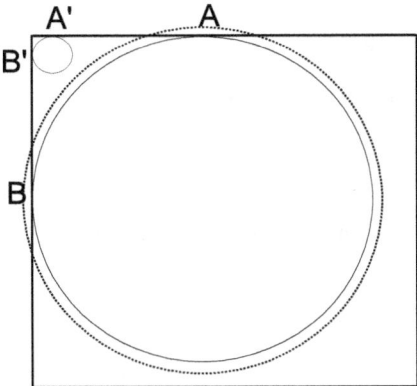

Figure 8: Two ordered elements in T(S) must share the same contact points (A,B). Furthermore for every disk D(A,B) of T(S) there exists an other disk D'(A',B') of T(S) such that $D \cap D' = \emptyset$.

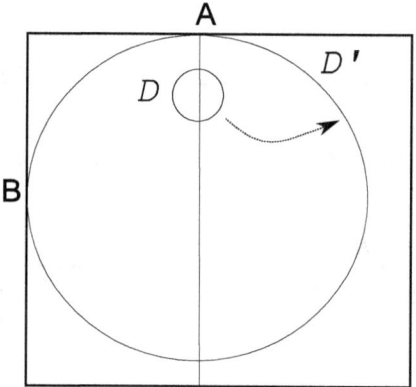

Figure 9: Mapping $D \to \varphi(D)$.

19/828

Subject: Hamel basis

Prove, using Zorn's lemma that every vector space has a basis.

SOLUTION:

We consider a vector space E such as $E \neq \{0\}$ and let us call $(S(E), \subset)$ the set of linearly independent vectors that belongs to E, partially ordered by inclusion. $S(E)$ is non empty because $E \neq \{0\}$.

If $\{S_i\}_{i \in I}$ is a totally ordered subset of $S(E)$, then $\bigcup_{i \in I} S_i$ is a set of non empty linearly independent vectors and this is an upper bound for $\{S_i\}_{i \in I}$.

This proves that $S(E)$ is inductively ordered.

By applying Zorn's lemma, we can see that $S(E)$ has at least one maximal element $\mathscr{H}$. $\mathscr{H}$ is a set of linearly independent vectors.

Let F be the subspace of E generated by $\mathscr{H}$. If we can find $\mathbf{x} \in F - E$ and $\mathbf{x} \neq 0$, then $\mathscr{H} \cup \{\mathbf{x}\}$ is linearly independent and it is strictly superior to $\mathscr{H}$, which is not possible. We conclude that $\mathscr{H}$ generates E, that is to say $\mathscr{H}$ is a basis of E.

This basis $\mathscr{H}$ is called a *Hamel basis*: a set of linearly independent vectors in a vector space E whose linear envelope is E.

Notes

[i]Hamel basis are named after the mathematician Georg Hamel who first used them in 1905 to describe the discontinuous (e.g. non-trivial) solutions of the Cauchy functional equation $f(x+y) = f(x) + f(y)$ (additive homomorphism from $(\mathbb{R}, +)$ into itself).

The original basis was the infinite basis $\{e_i\}_{i \in I}$ of $\mathbb{R}$, where $\mathbb{R}$ is seen as a $\mathbb{Q}$-vector space (so that every $x \in \mathbb{R}$ has a unique finite decomposition $x = \sum_{i=1,\ldots,n} q_i e_i, \; q_i \in \mathbb{Q}$).

From an historical point of view, it was an argument in order to introduce the axiom of choice in mathematics (we cannot prove the existence of such basis without the axiom of choice or one of its equivalent forms such as the Zorn's lemma).

A Hamel basis is then neither more nor less than the "classical" basis of a vector space but this denomination fits better when the vector space has infinite dimension and when the basis is not easily describable (such as some of the original Hamel basis of $\mathbb{R}$ over $\mathbb{Q}$ who lies somewhere inside the Cantor ternary set).

20/828

Subject: Equipotence relations between two sets A and B.

Show, with the help of Zermelo's theorem that, for any two sets A and B, there exist either a biunivocal application from A into a part of B, or a biunivocal application from B into a part of A.

SOLUTION:

a) With Zorn's lemma: the solution can be achieved through a very simple idea: we consider the family $\Lambda(A,B)$ of pairs (C,f) such that $C \subset A$ and $f : C \to B$ is an injective mapping. We define an order over these pairs by making:

$$(C,f) \leq (C',f') \text{ if} C \subset C' \text{ and } f'|_C = f.$$

Then $\Lambda(A,B)$ is not empty because $f_\emptyset$ defined on $\{\emptyset\}$ by $f_\emptyset(\{\emptyset\}) = \{\emptyset\}$ is such that $f_\emptyset$ belongs to it (or alternatively, we can define f_{ab} where $a \in A$, $b \in B$ and $f_{ab}(a) = b$, then $f_{ab} \in \Lambda(A,B)$).

If $S = (C_\alpha, f_\alpha)_{\alpha U}$ is a subset of $\Lambda(A,B)$ then we can define an injective mapping $\tilde{f}$ over $\bigcup_{\alpha \in U} C_\alpha$ by making $\tilde{f}|_{C_\alpha} = f_\alpha$ and then we see that $(\bigcup_{\alpha \in U} C_\alpha, \tilde{f})$ is an upper bound for S.

From Zorn's lemma this means that we can find a maximal element (C^m, f^m) in $\Lambda(A,B)$. Then either $C^m = A$, either $f^m(C^m) = B$ for if both does not hold, that is to say if we can find $a \in A - C^m$ and $b \in B - f^m(C^m)$ we can define $f^{m'}$ on $C^m \sqcup \{a\}$ by $f^{m'}|_B = f^m$ and $f^{m'}(a) = b$. But this would mean that $(C^m \sqcup \{a\}, f^{m'}) > (C^m, f^m)$ which is impossible. Then we have either $card(A) \leq card(B)$ or $card(B) \leq card(A)$.

b) With Zermelo's theorem: The process is exactly the same as in Exercise 15.b): we provide A and B with a total order, then we build a strictly increasing function $\varphi : A \to B$ by transfinite induction and then we distinguish the following cases:

1) both A and B have a maximal (= maximum) element,

2) neither A or B has a maximum,

3) A has a maximum but not B (and the symmetric case).

In all the cases, we see that either A and B are equivalent, or we can build a injection from one set into an other.

21/828 ¶

Subject: Equivalence between Zorn's lemma and Zermelo's theorem

a) Deduce Zermelo's theorem from Zorn's lemma ;

b) Prove Zorn's lemma from Zermelo's theorem .

SOLUTION:

a) *Zorn implies Zermelo*. Let A be a non-empty set. We consider the set S formed with the couples (X, ω) where $X \supset A$ and ω is a well-ordering of A. Obviously S is non empty for if $a \in A$, then $\{a\}$ provided with the trivial order is an element of S.

We define an order over S by: $(X, \omega) \leq (X', \omega')$ if:

1) $X \supset X'$,

2) the restriction of the well-order ω' to X is the well-order ω,

3) X is equal to $X'(m)$ (that is to say there exist a monotonous (strictly increasing) bijection between (X, ω) and $(X'(m), \omega')$).

S is inductively ordered : any totally non-empty ordered subsets of S has an upper bound in S. Indeed if we consider the family $\{X_\alpha, \omega_\alpha\}_{\alpha \in A}$ of totally ordered subsets of S. We consider $X = \bigcup_{\alpha \in A} X_\alpha$.

If $(x, y) \in X$ then we can find α such that $(x, y) \in X_\alpha$ (indeed, $\exists \alpha', \alpha''/x \in X_{\alpha'}, y \in X_{\alpha''}$ and either $X_{\alpha'} \subset X_{\alpha''}$, or $X_{\alpha''} \subset X_{\alpha'}$) so that $x \leq_{\omega_\alpha} y$ [41] is well defined regarding the induced order ω_α on X_α.

---

[41] We will note $\leq_{\omega_\alpha}$ the ω_α-order over X_α

If $(x,y) \in X_\beta$ with $X_\beta \subset X_\alpha$ or $X_\alpha \subset X_\beta$ then $x \leq_{\omega_\alpha} y$ is compatible with $x \leq_{\omega_\beta} y$ so that finally we may define an order over X. Let v denote this order. We want to show that v is a well-order of X. Indeed if Y is a non-empty subset of X, we can find α such that $(X_\alpha \cap Y) \neq \emptyset$ and let us define more generally, for all alpha's, $Y_\alpha = (X_\alpha \cap Y)$. Then: $Y = \bigcup_{\alpha \in A} Y_\alpha$.

Let y_α be the minimum element of Y_α (this element exists since $v|_{X_\alpha \cap Y} = \omega_\alpha$, that is to say, is a well-order of X_α). We suppose that $Y_\alpha \subset Y_{\alpha'}$ then we can find $m_{\alpha'}$ in $Y_{\alpha'}$ such that $Y_{\alpha'} = Y_{\alpha'}(m_{\alpha'})$.

We let y_α be the minimum of Y_α, then y_α is also the minimum of $Y_{\alpha'}$ so that we may define $y_m = y_\alpha$ independently of α and we see that y_m is the minimum of Y.

From this it follows that (X, v) is well-ordered and that X [42] is an upper bound for the set $\{X_\alpha\}_{\alpha \in A}$ which shows that the set S is inductively ordered.

We can therefore apply Zorn's lemma and claim that S has a maximal element (X^m, ω^m).

If $X^m \neq A$ then we can find $b \in A - X^m$ and from this we can build a well-ordered set in S: $(X^{m'}, \omega^{m'})$ by taking $X^{m'} = X^m \cup \{b\}$ and $\omega^{m'}|_{X^m} = \omega^m$, and defining b as the maximum in $X^{m'}$ so that $X^m = X^{m'}(b)$. This means that $(X^{m'}, \omega^{m'}) > (X^m, \omega^m)$ which is impossible since (X^m, ω^m) is a maximal element in S.

But this shows that we necessarily have $X^m = A$ and thus we can therefore provide A with a well-order, which is precisely Zermelo's theorem.

b)*Zermelo implies Zorn.*We consider the set X verifying the conditions of Zorn's Lemma - That is to say that X is inductively ordered: every subset S of X which is totally ordered has an upper bound (in X). Our target is to prove that this will imply that X has a maximal element.

For this, we consider, as suggested in the hint, $Y = P(X)$, the power set of X. Following Zermelo's theorem, we can well-order Y.

Let us suppose that y_m is the minimum element of Y, if $[y_m, y]$ denote $Y(y) \cup \{y\}$ ($Y(y)$ may also more conveniently refer to $[y_m, y)$) we build a set of functions φ^y from $[y_m, y]$ into X by the following rule:

we define $\varphi^{y_m}(y_m) = x_m$ where x_m is any point of X.

---

[42] As defined previously by $X = \bigcup_{\alpha \in A} X_\alpha$

If we have defined φ^y over $[y_m, y]$ for all the $y \in [y_m, y_0)$ then we may define φ^{y_0} as the function from $[y_m, y_0]$ into X such that $\varphi^{y_0}|_{[y_m, y]} = \varphi^y, \forall y \in [y_m, y_0)$ and $\varphi^{y_0}(y_0)$ as an element x_0 that is superior to all the elements in $\bigcup_{y \in [y_m, y_0)} \varphi^y([y_m, y]) = \varphi^y([y_m, y_0))$. Such an element always exist since X has no maximal element.

In order to prove that the definition of the functions φ^y is possible and gives birth to a function φ over the whole of Y to X. We recall here the principle of transfinite induction.

Let $(E, \leq)$ be a well-ordered set, if we have:

$0 = \min(E)$,

$U \subset E$, such that the following properties hold:

1) $0 \in U$,

2) $(\forall \alpha < x, \alpha \in U) \implies x \in U$,

then $U = E$.

The proof is rather straightforward: if $U \neq E$, then $V = E - U \neq \emptyset$ and $V \subset E$. We let $v = \min(V)$ and we see that $\forall \alpha < v$, we have $\alpha \notin V$ will imply that $\alpha \in U$, so that by 2) we must have $v \in U$ but this is impossible for $v \in V$ and $V \cap U = \emptyset$.

Then if we define here U as the set of $y \in Y$ such that φ^y is defined, we see that $y_m \in U$ and that $(\forall y < y_0, y \in U) \implies (y_0 \in U)$ so that by transfinite induction, we have $U = Y$ and φ^y is defined $\forall y \in Y$.

We then define φ over $\bigcup_{y \in Y}[y_m, y] = Y$ by:

$\varphi|_{[y_m, y]} = \varphi^y$ (or alternatively by $\varphi = \prod_{y \in Y} \varphi^y(y)$).

By construction φ is a strictly increasing bijective mapping from Y into X so that $card(Y) \leq card(X)$, and by Cantor's theorem (see [?]), we know that $Y = P(X) = 2^X$ has a strictly greater cardinality than X so this is impossible (note that we could have used instead of $P(X)$ any set Y of cardinality strictly greater than X).

The conclusion is that X must have a maximal element, what implies Zorn's Lemma.

22/828 ¶

Subject: Building of an order relation from a partial order

a) Show that, in a finite set X, every partial order relationship **R** is contained in an order relationship $\bar{\mathbf{R}}$;

b) Show that this is still true for infinite sets.

SOLUTION:

a) X is finite so that there exists a minimal element $x_1 \in X$. We iterate this process and from $X \setminus x_1$ we get a minimal element $x_2, \ldots, X \setminus \{x_1, \ldots, x_n\}$ is finite so we can find a minimal element x_{n+1}, etc...

We will exhaust the set X for a $n = N$, then we have totally ordered X in $\{x_1, x_2, \ldots, x_N\}$: if x_i and x_j are two elements of X with $i < j$, then we have:

$$x_i < x_{i+1} < \ldots < x_j.$$

From this result we get that $x_i < x_j$, so that any two elements can be compared.

b) Infinite case:

Solution 1:

We build $(\Lambda(X), \prec)$ the set of the partial order relationships over X ordered by the following relationship:

$$\forall(\Lambda_1, \Lambda_2) \in \Lambda(X)^2 , \ \Lambda_1 \prec \Lambda_2 \Leftrightarrow (\forall(x,y) \in X^2) \ x\Lambda_1 y \implies x\Lambda_2 y.$$

We consider an ordered subset $\Delta(X)$ of $\Lambda(X)$.

We define a partial order $\widetilde{\Lambda} = \bigsqcup_{\Lambda \in \Delta(X)} \Lambda$ by:

$$x\widetilde{\Lambda}y \Leftrightarrow \exists \Lambda \in \Delta(X), x\Lambda y.$$

Reflexivity: We have obviously $x\widetilde{\Lambda}x$.

Symmetry: If we have $x\widetilde{\Lambda}y$ and $y\widetilde{\Lambda}x$, then we can find Λ_1 and Λ_2 such that $x\Lambda_1 y$ and $y\Lambda_2 x$. If we have, $\Lambda_1 \preceq \Lambda_2$ (resp $\Lambda_2 \succeq \Lambda_1$) then we have:

$$x\Lambda_2 y, y\Lambda_2 x \ (\text{ resp } x\Lambda_1 y, y\Lambda_1 x).$$

So that in both cases: $x = y$.

Transitivity: If we have $x\widetilde{\Lambda}y$ and $y\widetilde{\Lambda}z$, then we can find Λ_1 and Λ_2 such that $x\Lambda_1 y$ and $y\Lambda_2 z$. If we have, $\Lambda_1 \preceq \Lambda_2$ (resp $\Lambda_2 \succeq \Lambda_1$) then we have:

$$x\Lambda_2 z \ (\text{ resp } x\Lambda_1 z).$$

So that in both cases: $x\widetilde{\Lambda}z$.

$\widetilde{\Lambda}$ is an upper bound element for $\Delta(X)$ so that the conditions of Zorn's lemma are met: the set $\Lambda(X)$ contains a maximal element Λ_{max} and we claim that this element must be a non-partial order relation.

Indeed, if we could find x and y in X such that x and y could not be compared by Λ_{max}, then we could define a new order Λ'_{max} such that $x\Lambda'_{max}y$ and one would have:

$$\Lambda_{max} \prec \Lambda'_{max}.$$

Which is impossible, given the definition of Λ_{max} as the maximal element of $\Lambda(X)$.

Solution 2:

We embed X into $P(X)$ using the application μ defined by:

$$X \rightarrow P(X),$$
$$x \rightarrow \mu(x) = \{y \in X, y \leq x\}.$$

We can use Zermelo's theorem to provide a well order to X, let us call $\prec$ this well-order.

We also note that we can define any element Z of $P(X)$ by its characteristic function χ_Z and, equivalently, by its "coordinates" over X, that is to say by the product:

$$\prod_{x \in X} \chi_Z(x).$$

So that the association $Z \rightarrow (\chi_Z(x))_{x \in X}$ span an equivalence[43]:

$$P(X) \approx \prod_X \{0, 1\}.$$

If we use *Exercise 13*, we see that we can define a lexicographical order over any set of the type: $\prod_{\alpha \in A} X_\alpha$, where A is well ordered and where X_α is a non-void ordered set for each α.

Then, here, we can take $A = (X, \prec)$ as the well-ordered set and $X_\alpha = \{0, 1\}$. We will then get a lexicographical order over $\prod_X \{0, 1\} = 2^X$. Let us call that order: $\lhd$.

We can transport this lexicographical order over $P(X)$ because of the equivalence between $P(X)$ and 2^X.

The order $\lhd$ over $P(X)$ contains the inclusion $\subset$ because, if we have $Y \subset Z$, then

$$Y \approx \prod_{x \in X} \chi_Y(x),$$
$$Z \approx \prod_{x \in X} \chi_Z(x).$$

---

[43]Two ordered set being equivalent in that context if we can find a strictly increasing bijective mapping between them

We know that:

$$\chi_Y(x) = 1 \implies \chi_Z(x) = 1.$$

So that when $\chi_Y(x) \neq \chi_Z(x)$, we will have $\chi_Y(x) = 0$ and $\chi_Z(x) = 1$ and:

$$Y \lhd Z.$$

Now we define the order $\ll$ over X as $\mu^{-1}(\lhd)$ (μ^{-1} being defined over $\mu(X)$), that is to say $x \ll x'$ if $\mu(x) \lhd \mu(x')$. Obviously, this new order, $\ll$ is a non-partial order. We still have to prove that it is an extension of our original partial order $<$, e.g. that $\ll$ contains $<$:

$$x < y \implies \mu(x) \subset \mu(y) \implies \mu(x) \lhd \mu(y) \implies x \ll y.$$

This shows that we can extend the partial order $<$ over an infinite set X into a total order $\ll$.

23/828 ¶

Subject: Transcendence basis of ℂ over ℚ

Show that the field of complex ℂ is isomorph to the algebraic closure of the field of rational functions with coefficients in ℚ of a continuum of algebraically independent variables.

warning:
One should read:
*(...) algebraic closure of the field of rational **expressions** (...)*
*a continuum of algebraically independent **elements** from ℂ.*

prerequisite:

We recall that some elements $\{x_\alpha\}$ over a field K are said to be algebraically independent if there is no non-trivial polynomial equation binding them together, e.g. for every $N \geq 0$, there is no polynomial $P \neq 0$ in $K[X_1, \ldots, X_N]$ such that $P(x_{\alpha_{i_1}}, \ldots, x_{\alpha_{i_N}}) = 0$.

We recall that, given a subfield K' of a field K, we say that $x \in K$ is algebraic over K' if it is the root of a non-trivial polynomial from $K'[X]$. The set of all the algebraic elements in K form the algebraic closure of K' in K.

discussion:

What is required is to find a family Ω of algebraically independent complexes z_α over ℚ and to form the field $\mathbb{Q}((z_\alpha)_{\alpha \in \Omega})$ of (finite) rational expressions of the z_α's with rational coefficients e.g. the field defined by the values:

$$R(z_{i_1},\ldots,z_{i_N}) = \frac{\sum_{j=1}^{j=N} a_{i_j} z_{i_j}^{m_{i_j}}}{\sum_{j=1}^{j=N} b_{i_j} z_{i_j}^{n_{i_j}}}, a_{i_j}, b_{i_j} \in \mathbb{Q}, m_{i_j}, n_{i_j} \in \mathbb{N}.$$

And then we must prove that $\mathbb{C}$ is the algebraic closure of $\mathbb{R}((z_\alpha)_{\alpha \in \Omega})$.

$\mathbb{Q}((z_\alpha)_{\alpha \in \Omega})$ is a subfield of $\mathbb{C}$ (this is trivially verified from its definition and from the fact that the numerator will never vanish because all the z_α are algebraically independent together), it is *not* a ring of polynomials or rational functions (the problem statement is a bit confusing). This is the smallest field containing $\mathbb{Q}$ and the z_α's.

We could first try to see $\mathbb{C}$ as a $\mathbb{Q}$-vector space and consider a Hamel basis of $\mathbb{C}$ (see ex 19), then we will get a family of complexes that are linearly independent over $\mathbb{Q}$ and that will span $\mathbb{C}$, so the algebraic closure of the sets of rational expression made with them will be the algebraic closure of $\mathbb{C}$ that is to say $\mathbb{C}$ itself. The problem is that this family is not necessarily algebraically independent over $\mathbb{Q}$ (e.g. it is irrational but not necessarily transcendent). We need to find a similar family but that will be, more generally, such that all elements are algebraically independent over $\mathbb{C}$.

solution:

We consider a transcendence base of $\mathbb{C}$ over $\mathbb{Q}$: if S is the set of algebraically independent complexes over $\mathbb{Q}$, then S is inductively ordered by inclusion and one can apply Zorn's lemma to S and produce at least one maximal family Ω. Ω is a transcendence base of $\mathbb{C}$ over $\mathbb{Q}$.

We form the field L=$\mathbb{Q}(\Omega)$ as described in the discussion. For any $z \in \mathbb{C}$-$\mathbb{Q}(\Omega)$, we must be able to find one polynomial $P \in \mathbb{Q}[X_1,\ldots,X_N]$ such that $P(z,z_{\alpha_1},\ldots,z_{\alpha_{N-1}}) = 0$ otherwise $\Omega \cup \{z\}$ would be a set of algebraically independent complexes strictly containing Ω, which is not possible. We consider $Q \in L[X]$ defined by $Q(X) = P(X,z_{\alpha_1},\ldots,z_{\alpha_{N-1}})$ and we see that $Q(z) = 0$.

So $\mathbb{C}$ is included in cl(L) (cl being the algebraic closure).

Since $\mathbb{C}$ is known to be such that cl($\mathbb{C}$)=$\mathbb{C}$, we have: cl(L)=$\mathbb{C}$.

We still need to prove that Ω is a continuum.

If we suppose that Ω is enumerable, we can order the z_i's.

We consider a sequence q in $\mathbb{Q}^\infty$ (a sequence in $\mathbb{N}^\infty$ would do well also) that converge to 0 and we build the polynomial $P = P_q$ defined by:

$$P_q(X) = \sum_{i \geq 0, a_i \geq 0} q_{a_0, a_1, \dots, a_p, \dots} [z_1^{a_1} \dots z_p^{a_p} \dots] X^{a_0}$$

or:

$$P_q(X) = \sum_n (\sum_{i \geq 1, a_i \geq 0} q_{a_0, a_1, \dots, a_p, \dots} [z_1^{a_1} \dots z_p^{a_p} \dots]) X^n.$$

We note that $q \to P_q$ is an injection because, by looking at the values of $P_q(x)$ when $x = 1$, we see easily that $q \neq q'$ imply that $P_q(1) \neq P_{q'}(1)$ otherwise this would create an algebraic dependence in Ω. However that is not important for our purpose.

For every sequence q, let us note Δ_q the roots of P_q in $\mathbb{C}$. We know that $card(\Delta_q) = deg(P_q) < \infty$. When q describes the set of rational sequences converging toward 0 [44], Δ_q will recover $\mathbb{C}$ so that:

$$\bigcup \Delta_q = \mathbb{C}.$$

That means:

$$\sum_{q \in \mathbb{Q}^\infty, q \to 0} card(\Delta_q) \geq card(\mathbb{C}).$$

$\mathbb{N}$ is equipotent to $\mathbb{N}^2$ who is itself equipotent to $\mathbb{Q}$. We can establish a bijection between the set $\{q_1, \dots, q_n, 0 \dots 0 \dots q_i \in \mathbb{Q}\}$ and $\mathbb{N}^n$ so that the set of rational sequences who contains only a finite amount of non null terms is equipotent to $\bigcup_{n=o}^{\infty} \mathbb{N}^n$.

If α represents a sequence q, we note Δ_α for Δ_q and:

$$\sum_{q \in \mathbb{Q}^\infty, q \to 0} card(\Delta_q) = \sum_{\alpha \in \bigcup_{n=o}^{\infty} \mathbb{N}^n} card(\Delta_\alpha).$$

---

[44]E.g. the set of rational sequences which have only finite non-null terms

An enumerable union of enumerable sets is also enumerable, so that we can find a bijection f between: $\bigcup_{n=0}^{\infty} \mathbb{N}^n$ and $\mathbb{N}$. If we note again $n = f(\alpha)$ and Δ_n for Δ_α, then we have:

$$\sum_{q \in \mathbb{Q}^\infty, q \to 0} card(\Delta_q) = \sum_{n \in \mathbb{N}} card(\Delta_n)$$

and we know that $card(\Delta_n)$ is finite for all $n \in \mathbb{N}$ and so as, the cardinal of a disjunctive enumerable union of finite sets has the power of $card(\mathbb{N}) = \aleph_0, \sum_{n \in \mathbb{N}} card(\Delta_n) = \aleph_0$.

But this is impossible since that would mean that $\mathbb{C}$ would not a continuum.

Notes

[i] An enumerable union of enumerable sets is enumerable, $\mathbb{N}^n$ is equipotent to $\mathbb{N}$.

Let $\{D_n\}_n$ be a an enumerable set of enumerable sets D_n. Then $\bigcup_{n \in \mathbb{N}} D_n$ is equivalent to the infinite matrix $\{d_{i,j}\}_{(i,j) \in \mathbb{N}^2}$ where $d_{i,j}$ is the jth element of D_i. We can the define φ as the following application:

$$\bigcup_{n \in \mathbb{N}} D_n \to \mathbb{N}^2,$$

$$d_{i,j} \to \varphi(d_{i,j}) = (i,j).$$

We see that $card(\bigcup_{n \in \mathbb{N}} D_n) \leq card(\mathbb{N}^2)$. But $\mathbb{N}^2$ is equipotent to $\mathbb{N}$. This is a well-known result from basic set theory and we recall here two ways of proving it:

a) we can consider the bijection $f : \mathbb{N}^2 \to \mathbb{N}$ defined by $f(m,n) = \frac{n^2+3n-2m+2}{2}$. This function is a "diagonal enumeration" of the points in $\mathbb{N}^2$. Indeed the number S of points in the triangle of coordinates $(0,0); (m,0); (0,m)$ is $(m+1) + m + \ldots + 1 = \frac{(m+1)(m+2)}{2}$. If we define an order on $\mathbb{N}^2$ by $(i,j) < (i',j') \Leftrightarrow ((i+i' < j+j')$ or $(i+i' = j+j'$ and $j < j'))$ then we have a (non-partial) "diagonal-lexicographical" order over $\mathbb{N}^2$ and the amount of points between $(0,0)$ and (m,n) is $= S - n$ that is to say $\frac{n^2+3n-2m+2}{2}$.

b) we can use the mapping:

$$\mathbb{N}^2 \to \mathbb{N},$$

$$(m,n) \to 2^m 3^n.$$

This mapping is an injection and as we can also consider the canonical injection $\mathbb{N} \to \mathbb{N}^2$ we can use the Cantor-Bernstein theorem to claim that there is a bijection between $\mathbb{N}$ and $\mathbb{N}^2$.

So that $card(\bigcup_{n \in \mathbb{N}} D_n) \leq card(\mathbb{N})$. We also have $card(\bigcup_{n \in \mathbb{N}} D_n) \geq card(\mathbb{N})$ and we can conclude that: $card(\bigcup_{n \in \mathbb{N}} D_n) = card(\mathbb{N})$.

Incidentally, by immediate recursion, we check that $\mathbb{N}^n$ is equipotent to $\mathbb{N}$.

Exercises 24 to 46
Completions

"Do not let him open a book of mathematics, nor writing a figure before he achieved his litterary studies".

Advice given by the French mathematician Lagrange to Augustin Cauchy's father regarding the education of his son, showing at the time precocious capacities in mathematics. Augustin Louis, baron Cauchy (1789-1857) was the first to define the criterion for fundamental sequences (in: Jean Baptiste Biot, Mélanges scientifiques et littéraires).

24/828 ¶

Subject: The Shrinking Balls Theorem

a) Show that in a complete space (X, d_X) the *Shrinking Balls Theorem* apply:

Let B_n be a sequence of closed balls in a metric space X such that:
1) $B_1 \supset B_2 \supset \ldots \supset B_n \supset \ldots$

2) If ρ_n is the radius of the ball B_n, then $\rho_n \to 0$, $n \to \infty$.

Then:

$$\bigcap_{n=1}^{\infty} B_n = \{x\}, x \in X.$$

b) Show that if the shrinking ball theorem is true in a metric space X, then the space is complete.

SOLUTION:

a) We consider $\{x_n\}_n$ the sequence of the centers of the balls B_n:

$$(\forall (p.q) \in \mathbb{N}^2), p < q \implies B_p \supset B_q \implies d_X(x_p, x_q) \leq 2\rho_p.$$

So that $\{x_n\}_n$ is a Cauchy sequence and therefore must converge in X since the space is complete. Let $x = \lim_{n \to \infty} x_n$. We have:

$$\forall n > 0, x \in B_n.$$

So that:

$$x \in \bigcap_n B_n.$$

If we can find $y \in X$ such that $y \neq x$ and $y \in \bigcap_n B_n$ then we would have $0 < d_X(x,y) < 2\rho_n, \forall n > 0$ and this is impossible since ρ_n tends towards 0, we would have $d_X(x,y) = 0$ and hence $x = y$.

To conclude, we have:

$$\{x\} = \bigcap_n B_n.$$

b) Let us suppose that the shrinking ball theorem is true in a metric space X. Let $\{x_n\}_n$ be a Cauchy sequence.

We can find a subsequence $x_{\varphi(k)} = x_{n_k}$ such that:

$$\forall l > k, x_{n_l} \in B(x_{n_k}, \frac{1}{2^k})\ ^{45}.$$

We can find x_{n_1} such that $B(x_{n_1}, \frac{1}{2})$ contains infinitely many x_i's. Let us call $\{x_n\}^{(1)}$ this infinite set contained in $B(x_{n_1}, \frac{1}{2})$. From $\{x_n\}^{(1)}$, we can find x_{n_2} such that $B(x_{n_2}, \frac{1}{4})$ contains infinitely many elements from $\{x_n\}^{(2)}$, we call $\{x_n\}^{(2)}$ this infinite set contained in $B(x_{n_2}, \frac{1}{4})$, etc...

We define the balls $\widetilde{B_k}$ by: $\widetilde{B_k} = B(x_{n_k}, \frac{1}{2^{k-1}})$.

We consider $\widetilde{B_k}$ and $\widetilde{B}_{k+1}$ and let $x \in \widetilde{B}_{k+1}$ then $d_X(x, x_{n_{k+1}}) \leq \frac{1}{2^k}$.

We also have $d_X(x_{n_k}, x_{n_{k+1}}) \leq \frac{1}{2^k}$ because $x_{k+1} \in \widetilde{B}_k$ so that:

$$\begin{aligned}
d_X(x, x_{n_k}) &\leq d_X(x, x_{n_{k+1}}) + d_X(x_{n_k}, x_{n_{k+1}}), \\
&\leq \frac{1}{2^k} + \frac{1}{2^k}, \\
&= \frac{1}{2^{k-1}}.
\end{aligned}$$

---

[45] We note $B(x,r) = \{y \in X, d_X(x,y) \leq r\}$

So that, finally, $x \in \widetilde{B}_{k+1} \Rightarrow x \in \widetilde{B}_k$ and $\widetilde{B}_{k+1} \subset \widetilde{B}_k$.

The radius ρ_n of the balls tends toward 0 so that the balls are shrinking.

We use the shrinking Balls theorem (that we just proved in a)) and we see that the balls intersect in:

$$\bigcap_k B_k = \{x\}, x \in X.$$

This implies that x is the limit of the sequence $\{x_n\}_n$, since for all $\varepsilon > 0$, we can find k such that $\frac{1}{2^k} < \varepsilon$ and x is the limit of the sequence $\{x_{n_k}\}_k$ when $k \to \infty$, and then, since $\{x_n\}_n$ is a Cauchy sequence, we must have $x = lim_{n \to \infty} x_n$. This proves that X is a complete space.

25/828 $\mathscr{B}$

Subject: Uniformly continuous function on a complete space

Show that any uniformly continuous numeric function on a metric space (X, d_X) may be extended in an unique way in a continuous function on the completion of X and that this extension is uniformly continuous.

SOLUTION:

Let $\{x_n\}$ be a Cauchy sequence in X:

f is uniformly continuous so that $\forall \varepsilon > 0$, we can find $\alpha > 0$ such that:

$$\forall (p,q) \, , \, d_X(x_p, x_q) < \alpha \implies |f(x_p) - f(x_q)| < \varepsilon.$$

But we also know, since $\{x_n\}_n$ is a Cauchy sequence, that $\forall \alpha > 0$, we can find $N > 0$ such that:

$$(p,q) > N \implies d_X(x_p, x_q) < \alpha.$$

So that, by combining the two results:

$$\forall \varepsilon > 0 \, , \, \exists N > 0, (p,q) > N \implies |f(x_p) - f(x_q)| < \varepsilon.$$

This proves that $\{f(x_n)\}$ is also a Cauchy sequence and therefore, $\{f(x_n)\}_n$, being a Cauchy sequence will converge to a $z \in \mathbb{R}$ since $\mathbb{R}$ is complete.

If $x_n \to x \in \widetilde{X}$ [46] then we may extend f to $\widetilde{X}$ by:

---

[46] $\widetilde{X}$ being the completion of X

$$\tilde{f}(x) = z = \lim_{n \to \infty} f(x_n).$$

Rigourously, we must check the well-definition of the extension.

It is needed to check that if $x_n \to x$ and $y_n \to y$ then $\lim f(x_n) = \lim f(y_n)$: in this case, we have $d_X(x_n, y_n) \to 0$ and this implies $|f(x_n) - f(y_n)| \to 0$.

The extension is obviously continuous. Let us show that it is uniformly continuous.

Let $(x,y) \in \tilde{X}$. If $x = \lim_{n \to \infty}(x_n)$ and $y = \lim_{n \to \infty}(y_n)$, $\{x_n\}_n$ and $\{y_n\}_n$ being Cauchy sequences in X, then we know from the uniform continuity of f that: $(\forall \varepsilon), \exists \alpha, \forall (u,v) \in X^2, d_X(u,v) < \alpha \Rightarrow |f(u) - f(v)| < \varepsilon$. If we make $u = x_n$ and $v = y_n$, then we get $d_X(x_n, y_n) < \alpha \Rightarrow |f(x_n) - f(y_n)| < \varepsilon$.

So that at the limit $n \to \infty$:

$$\forall \varepsilon > 0, \exists \alpha > 0, d_{\tilde{X}}(x,y) < \alpha \implies |\tilde{f}(x) - \tilde{f}(y)| < \varepsilon$$

and this proves the uniform continuity of $\tilde{f}$.

26/828

Subject: Completion of some metric spaces

Show that the following metric spaces are not complete and build their completions:

a) $\mathbb{R}$ provided with the distance $d(x,y) = |\arctan(x) - \arctan(y)|$;

b) $\mathbb{R}$ provided with the distance $d(x,y) = |\exp(x) - \exp(y)|$.

SOLUTION:

If f is an injection $\mathbb{R} \to \mathbb{R}$, then we may define d_f by $d_f(x,y) = |f(x) - f(y)|$. d_f is a distance since:
- $d_f(x,x) = 0$;
- $d_f(x,y) = 0 \Rightarrow f(x) = f(y) \Rightarrow x = y$;
- $d_f(x,y) + d_f(y,z) = |f(x) - f(y)| + |f(y) - f(z)| \geq |f(x) - f(y) + f(y) - f(z)| = d_f(x,z)$.

This proves the validity of the definition of the distances in a) and b).

a) The application $f : x \to \arctan(x)$ is an isometry of $(\mathbb{R},d)$ into $(] - \frac{\pi}{2}, \frac{\pi}{2}[, d_0)$ [47] since $d(x,y) = d_0(f(x), f(y))$, and we know that, from Exercise 31, *in a complete metric space, the completion of a subset is its closure*, the completion of $(] - \frac{\pi}{2}, \frac{\pi}{2}[, d_0)$ is $([- \frac{\pi}{2}, \frac{\pi}{2}], d_0)$ so that:

the completion of $(\mathbb{R}, d)$ is isometric to $([- \frac{\pi}{2}, \frac{\pi}{2}], d_0)$.

[47] d_0 being the ordinary distance

b) The application $g : x \to \exp(x)$ is also an isometry of $(\mathbb{R}, d)$ into $(]0, \infty[, d_0)$ so that, again, the completion of $(]0, \infty[, d_0)$ is $([0, \infty[, d_0)$ and:

$$\text{the completion of } (\mathbb{R}, d) \text{ is isometric to } ([0, \infty[, d_0).$$

Notes

[i] In the general case, if $d_f(x, y) = |f(x) - f(y)|$ where f is an injection from $\mathbb{R}$ into $\mathbb{R}$, then f is an isometry from $(\mathbb{R}, d_f)$ into $(f(\mathbb{R}), d_0)$ and the completion of $(\mathbb{R}, d_f)$ is isometric to $(\overline{f(\mathbb{R})}, d_f)$ where $\overline{f(\mathbb{R})}$ is the closure of $f(\mathbb{R})$ regarding the usual metric d_0 of $\mathbb{R}$.

27/828 $\mathscr{B}$

Subject: Completion of a set of intervals

Over the set $\Delta(\mathbb{R})$ of closed[48] intervals of $\mathbb{R}$, we define a distance by:

$$d([a,b],[c,d]) = |a-c| + |b-d|.$$

Show that this metric space is not complete and find its completion.

SOLUTION:

We must first check that d is a distance over $\Delta(\mathbb{R})$:

- $d([a,b]),[a,b]) = 0$;
- $d([a,b],[c,d]) = 0 \Rightarrow |a-c| + |b-d| = 0 \Rightarrow a=c$ and $b=d \Rightarrow [a,b] = [c,d]$;
- $d([a,b],[c,d]) + d([c,d],[e,f]) = |a-c| + |b-d| + |c-e| + |d-f| = (|a-c| + |c-e|) + (|b-d| + |d-f|) \geq |a-e| + |b-f| = d([a,b],[e,f])$.

Let us define the sequence $D_n = \{[a - \frac{1}{n}, a + \frac{1}{n}]\}_n$:

$d_\Delta(D_p, D_q) = 2|\frac{1}{q} - \frac{1}{p}|$ so that $\{D_n\}$ is a Cauchy sequence but converge towards a singleton $\{a\}$ that is not an element of $\Delta(\mathbb{R})$ and hence $\Delta(\mathbb{R})$ is not complete.

If we consider again a Cauchy sequence $D_n = [a_n, b_n]$ then the fact that:

$$d(D_p, D_q) = |a_p - a_q| + |b_p - b_q| \to 0$$

---

[48] Closed and bounded intervals

implies that $\{a_n\}$ and $\{b_n\}$ must be Cauchy sequences such that $a_n < b_n$ so that, because $\mathbb{R}$ is complete, $a_n \to a$ and $b_n \to b$. Therefore $a \leq b$ and:

1. $D_n \to [a,b]$ if $a < b$;

2. $D_n \to \{a\}$ if $a = b$.

The set of limits of Cauchy sequences is then being described by:

$$\Delta(\mathbb{R}) \sqcup \bigcup_{x \in \mathbb{R}} \{x\}.$$

Since the metric $\tilde{d}$ of the completed space $\Delta(\tilde{\mathbb{R}})$ is such that $\tilde{d}(\{x\},\{y\}) = 2|x-y|$, this implies that $x \neq y \Rightarrow \{x\} \neq \{y\}$ in $\Delta(\mathbb{R})$.

We conclude that

$$\Delta(\tilde{\mathbb{R}}) = \Delta(\mathbb{R}) \sqcup \bigcup_{x \in \mathbb{R}} \{x\}.$$

28/828

Subject: Completion of a set of intervals provided with symmetric difference

Over the set $\Delta(\mathbb{R})$ of closed intervals[49] of $\mathbb{R}$, we define the following distance as a symmetric difference:

$$d(\Delta_1, \Delta_2) = |\Delta_1| + |\Delta_2| - 2|\Delta_1 \cap \Delta_1|.$$

Show that this space is not complete and build its completion.

SOLUTION:

First, we need to check that d is well-defined as a distance:
$d(\Delta_1, \Delta_2) = |\Delta_1 \triangle \Delta_2|$, where $\triangle$ is the symmetric difference. This implies that:

1) $d(\Delta_1, \Delta_1) = 0$,

2) $d(\Delta_1, \Delta_2) = 0 \Rightarrow \Delta_1 \triangle \Delta_2 = \emptyset \Rightarrow \Delta_1 = \Delta_2$,

3) $d(\Delta_1, \Delta_2) + d(\Delta_1, \Delta_2) = |\Delta_1 \triangle \Delta_2| + |\Delta_2 \triangle \Delta_3| = |\Delta_1^{23}| + |\Delta_{13}^2| + |\Delta_2^{13}| + |\Delta_{23}^1| + |\Delta_2^{13}| + |\Delta_{12}^3| + |\Delta_3^{12}| + |\Delta_{13}^2|$ and since $|\Delta_1 \triangle \Delta_3| = |\Delta_1^{23}| + |\Delta_{12}^2| + |\Delta_3^{12}| + |\Delta_{23}^1|$, we have $d(\Delta_1, \Delta_2) + d(\Delta_1, \Delta_2) \geq d(\Delta_1, \Delta_3)$ (We have partitioned $\Delta_1 \cup \Delta_2 \cup \Delta_3$ see Fig. 10 for the details).

This proves that d is well-defined as a distance.
We consider a contractile sequence of intervals $\{\Delta_n\}$, that is to say a sequence such that:

---

[49] See Exercise 27

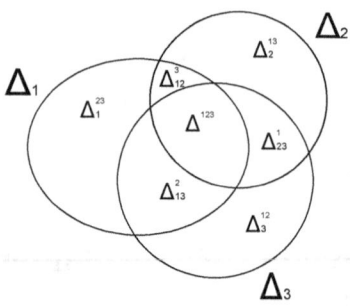

Figure 10: Partition of $\Delta_1 \cup \Delta_2 \cup \Delta_3$

$$\Delta_1 \supset \ldots \supset \Delta_n \supset \ldots \,, \ |\Delta_n| \to 0.$$

This sequence is a Cauchy sequence because, for $p < q$:

$$d(\Delta_p, \Delta_q) = |\Delta_p| + |\Delta_q| - 2|\Delta_p \cap \Delta_q| = |\Delta_p| - |\Delta_q| \leq |\Delta_p| \ ^{50}.$$

This contractile sequence tends towards a point $\{x\} = \bigcap_{n=1}^{n=\infty} \Delta_n$ and this point is not an element of $\Delta(\mathbb{R})$. This proves the non-completeness of $\Delta(\mathbb{R})$.

Furthermore, for two such points $\{x\}$ and $\{y\}$, we have $\tilde{d}(\{x\}, \{y\}) = |\{x\}| + |\{y\}| - 2|\{x\} \cap \{y\}| = 0$. So that $\{x\} = \{y\}$ in $(\Delta(\tilde{\mathbb{R}}), \tilde{d})$. We will name u this point.

From the hint, we need to prove that from any non contractile Cauchy sequence $\{\Delta_n\}_{n \geq 0}$ and *such that* $|\Delta_n|$ *does not converge towards* 0 [51], we can extract a subsequence $\{\Delta_{n_i}\}$ such that $\Delta_{n_i} \cap \Delta_{n_j} \neq \emptyset, \forall (i,j) \in \mathbb{N}^2$.

1) If $\{\Delta_n\}_{n \geq 0}$ is a Cauchy sequence such that $|\Delta_n| \to 0, n \to \infty$, then it is clear that $\tilde{d}(u, \Delta_n)$ tends toward 0 and hence $\{\Delta_n\}_{n \geq 0}$ will converge towards u.

---

[50] Since $|\Delta_p \cap \Delta_q| = |\Delta_q|$

[51] This last precision was absent from the original Russian version of TPFA

2) Let us suppose now that the length of the Δ_n does *not* tend toward 0. If we suppose that $\forall N \in \mathbb{N}$, we can find $(p,q) \in \mathbb{N}^2$ such that $(p,q) > N$ and $\Delta_p \cap \Delta_q = \emptyset$, then this means that $\forall \varepsilon > 0, \exists (p,q) \in \mathbb{N}^2, |\Delta_p| + |\Delta_q| < \varepsilon$ and this implies that the length of the intervals tends toward 0, which is impossible. We can therefore find a subsequence of $\{\Delta_n\}_{n \geq 0}$ such that any interval intersects the other one. Since the distance d is equal to the distance defined in the Exercise 27 in the case intervals intersects each other, this subsequence will converge towards a point $\{x\}, x \in \mathbb{R}$, that is to say will converge towards u.

We conclude that $\Delta(\tilde{\mathbb{R}}) = \Delta(\mathbb{R}) \cup u$.

29/828 ℬ

Subject: Completeness of a functional space

Show that the space $B(X)$ of all the bounded numeric functions over X provided with the distance:

$$d(f,g) = \sup_{x \in X} |f(x) - g(x)|$$

is a complete space.

SOLUTION:

If $\{f_n\}$ is a Cauchy sequence, then $\forall x, \{f_n(x)\}$ is also a Cauchy sequence in $\mathbb{R}$ and thus, tends toward a numeric value $f(x) = \lim_{n \to \infty} f_n(x)$: indeed, $\forall x, \forall (p,q), |f_p(x) - f_q(x)| \leq d(f_p, f_q)$.

The convergence $f_n \to f$ is uniform.

For $\varepsilon > 0$, $\exists N_\varepsilon$ so that, $\forall N > N_\varepsilon$, we can find a subset $\{n_k\}$ with:

$$n_1 = N, \forall x, \forall k > 0, |f_{n_k}(x) - f_{n_{k+1}}(x)| < d(f_{n_k}, f_{n_{k+1}}) < \varepsilon/2^k.$$

So that:

$$\forall \varepsilon > 0, \exists N_\varepsilon, \forall N > N_\varepsilon, |f(x) - f_N(x)| < \varepsilon.$$

The fact that f is bounded over X comes from:

$$|f(x)| \leq |f(x) - f_n(x)| + |f_n(x)|.$$

Since f_n converge uniformly toward f, for a $M > 0$, we can find N so that:

$$\forall n > N, (\forall x \in X), |f(x) - f_n(x)| < M$$

and since $f_n \in B(X)$, we have:

$$\sup_{x \in X} |f(x)| \leq M + \sup_{x \in X} |f_n(x)| = M + d(f_n, 0) \;\;^{52}$$

and this implies $f \in B(X)$.

We conclude that every Cauchy sequence in $B(X)$ converge towards an element of $B(X)$ so that $B(X)$ is complete.

---

[52] $d(f_n, 0)$ being the distance from f_n to the constant function $\mathbb{R} \to \mathbb{R}, x \to 0$

30/828 $\mathscr{B}$

Subject: Isometry between X and $B(X)$

Let (X,d) be a bounded metric space. Prove that $x \to d(x,.)$ is an isometry from X into $B(X)$.

SOLUTION:

Let $f_x \in B(X)$ be defined by $f_x(u) = d_X(x,u)$:

$$\forall (x,y) \in X^2, d_{B(X)}(f_x, f_y) = \sup_{u \in X} |f_x(u) - f_y(u)|,$$
$$= \sup_{u \in X} |d_X(x,u) - d_X(y,u)|,$$
$$\leq d_X(x,y) (\text{since } |d(x,u) - d(u,y)| \leq d(x,y)).$$

If $u = x$ or $u = y$, $|d_X(x,u) - d_X(y,u)| = d_X(x,y)$ so that:

$$d_{B(X)}(f_x, f_y) = d_X(x,y).$$

31/828 $\mathscr{B}$

Subject: A theorem of completeness for bounded spaces

X being a subset in a complete metric space Y, prove that:

a) X is complete if and only if it is closed;

b) The completion of X is equal to its closure in Y;

c) Deduce from *Exercises 29,30* and *a)* a theorem of completeness for bounded spaces.

SOLUTION:

a) b)Let $\{x_n\}$ be a Cauchy sequence in X. Then, if $\tilde{X}$ is the completion of X, x_n is convergent in $\tilde{X}$. Let $x \in \tilde{X}$ be the element $x = \lim_{n \to \infty} x_n \in \tilde{X}$, $x \in \overline{X}$ [53] since it is the limit of a sequence from X and we have:

$$\tilde{X} \subseteq \overline{X}.$$

Conversely, if $x \in \overline{X}$ then x is the limit of a sequence $\{x_n\}$ where $x_n \in X, \forall n > 0$. $\{x_n\}$ is convergent thus is a Cauchy sequence in X so that $x \in \tilde{X}$ and this leads to:

$$\tilde{X} \supseteq \overline{X}.$$

---

[53] $\overline{X}$ being the closure of X

So that, we conclude:

$$\widetilde{X} = \overline{X} \quad {}^{54}.$$

c) We know from *Exercise 29* that the space $(B(X),d)$ provided with
the distance $d(f,g) = \sup_{x \in X} |f(x) - g(x)|$ is complete and, using *Exercise
30*, we know that we can embed isometrically any bounded metric space X
into $B(X)$ using the correspondence $x \to d(x,.)$.

Then by using the results of a) and b), we deduce that *the completion of
any bounded metric space X is isometric to the adherence into $B(X)$ of its
image by the correspondence $x \to d(x,.)$.*

---

[54] It is neither nor less than the immediate application of the principle that Cauchy sequences are equivalent to convergent sequences in a complete space

32/828

Subject: Intersection of everywhere dense sets

Let (X,d) be a complete metric space and let $\{Y_i\}$ be a sequence of open subsets that are everywhere dense in X. Prove that $\bigcap_i Y_i$ is also everywhere dense in X.

SOLUTION:

We consider a point $x \in X$: for $\varepsilon > 0$ fixed, Y_0 is everywhere dense in X so that we can find $y_0 \in Y_0$ such that $|x - y_0| < \varepsilon/2$. Y_0 is open so we can find a closed ball $B_0 = B(y_0, \rho_0)$ such that $B_0 \subset Y_0$ and $\rho_0 < \varepsilon/2$. Y_1 is also everywhere dense in X so that we can find $y_1 \in B_0 \cap Y_1$. Again, Y_1 is open so we can find $B_1 = B(y_1, \rho_1)$ so that $B_1 \subset Y_1$. If needed, we can decrease ρ_1 so that $B_1 \subset B_0$ and $\rho_1 < \rho_0$. By iterating this process, if we already have:

1. $B_n \subset B_{n-1} \subset \ldots \subset B_0$;

2. $B_n = B(y_n, \rho_n), y_n \in Y_n$;

3. $\rho_n < \rho_{n-1}/2$;

4. $B_n \subset Y_n$.

Since Y_{n+1} is everywhere dense so that we can find $y_{n+1} \in Y_{n+1} \cap B_n$. Y_{n+1} is open and we can find $B_{n+1} = B(y_{n+1}, \rho_{n+1})$ so that $B_{n+1} \subset B_n$ and $\rho_{n+1} < \rho_n/2$ (again we can decrease ρ_{n+1} if needed), etc...

Finally we get a shrinking sequence of balls $\{B_n\}$ and by using the *Shrinking Balls Theorem* (see Exercise 24) in the complete space X, we have:

$$\bigcap_{n \geq 0} B_n = \{x_0\}, x_0 \in X.$$

We know that:

$$\bigcap_{n \geq 0} B_n \subset \bigcap_{n \geq 0} Y_n.$$

So that this implies:

$$x_0 \in \bigcap_{n \geq 0} Y_n.$$

And, we see that:

$$d(x, x_0) \leq d(x, y_0) + d(y_0, y_1) + \ldots d(y_{n-1}, y_n) + d(y_n, x_0),$$
$$\leq \varepsilon/2 + \varepsilon/4 + \ldots + \varepsilon/2^n + d(y_n, x_0).$$

If $n \to \infty$, $d(y_n, x_0) \to 0$ and $d(x, x_0) \leq \varepsilon$.

This proves that $\bigcap_{n \geq 0} Y_n$ is everywhere dense in X.

33/828

Subject: Properties of $\mathbb{R}\backslash\mathbb{Q}$

a) Show that the set of irrational numbers of $\mathbb{R}$ cannot be represented by the union of an enumerable number of closed sets.

b) Show that, on $\mathbb{R}$, there is no function that is continuous on all the rational points and discontinuous on all the irrational points.

SOLUTION:

a) Let us suppose that $\mathbb{R} - \mathbb{Q}$ can be written as the enumerable union of closed sets then $\mathbb{Q}$ could be written as the enumerable intersection of some open sets Γ_n:

$$\mathbb{Q} = \bigcap_{n=0}^{n=\infty} \Gamma_n.$$

Then we use the fact that $\mathbb{Q}$ is enumerable and we put: $\mathbb{Q} = \{r_n, n \geq 0\}$. We note that the sets Γ_n must be dense in $\mathbb{R}$ because they contain a dense subset of $\mathbb{R}$: $\mathbb{Q}$. Next we define the sets $\widetilde{\Gamma_n}$ by:

$$\widetilde{\Gamma_n} = \Gamma_n \backslash r_n.$$

The sets $\widetilde{\Gamma_n}$ are still dense and open and we have:

$$\bigcap_{n=0}^{n=\infty} \widetilde{\Gamma_n} = \emptyset.$$

But this is a contradiction with the result of *Exercise 32* that tells us that an enumerable intersection of everywhere dense sets is also everywhere dense. So that $\bigcap_{n=0}^{n=\infty} \widetilde{\Gamma}_n$ should be everywhere dense.

We conclude that we cannot represent $\mathbb{R} - \mathbb{Q}$ as the enumerable union of closed sets.

b) Let us define the oscillation of a function f in the interval $[a,b]$ by:

$$\omega_f([a,b]) = \sup_{a \le x \le y \le b} (|f(x) - f(y)|)$$

and the oscillation of f at the point x by:

$$\omega_f(x) = \lim_{\varepsilon \to 0} \omega_f[x - \varepsilon, x + \varepsilon].$$

We see that, $\forall c > 0$, the sets $F_c(f) = \{x \in X, \omega_f(x) \ge c\}$ are closed: if $x_0 \in (X \backslash F_c(f)) = \{x \in X, \omega_f(x) < c\}$, then we can also find $r > 0$ so that, for every $x \in [x_0 - r, x_0 + r]$, $\omega_f(x) < c$ (Otherwise, if we would have that $\forall r > 0, \exists x \in [x_0 - r, x_0 + r]$, $\omega_f(x) > c$ then we would have $\omega_f(x_0) \ge c$, which is impossible).

The set of points F_f where f is discontinuous is the set of points where the oscillation is > 0, that is to say:

$$F_f = \bigcup_{c>0} F_c(f).$$

We have $c < d \implies F_c \subset F_d$ so that we can describe F_f as the enumerable union of closed sets:

$$F_f = \bigcup_{n>0} F_c(\frac{1}{n}).$$

If we can find a function f that is continuous over $\mathbb{R}\backslash\mathbb{Q}$ and discontinuous over $\mathbb{Q}$, then we could represent $\mathbb{R}\backslash\mathbb{Q}$ as the the enumerable union of closed sets:

$$\mathbb{Q} = \bigcup_{n>0} F_c(\frac{1}{n}).$$

But, from a) we know that this is impossible and we conclude that we cannot find such a function.

34/828

Subject: Non-completeness of some polynomial spaces

Show that the polynomial spaces $(\mathbb{R}[X], d)$ are not complete for the following distances d:

a) $d_1(P, Q) = \max_{x \in [0,1]} |P(x) - Q(x)|$;

b) $d_2(P, Q) = \int_0^1 |P(x) - Q(x)| dx$;

c) $d_3(P, Q) = \sum_i |c_i|$ if $P(x) - Q(x) = \sum c_i x^i$.

SOLUTION:

The fact that d_1, d_2 and d_3 are distances is easy to check and we skip the details.

Now, we consider the sequence $\{P_N\}$ defined by:

$$P_N(x) = \sum_{n=0}^{n=N} (\frac{x}{2})^n.$$

a) We have, for $p < q$, $d_1(P_p, P_q) = \max_{x \in [0,1]} |\sum_{n=p+1}^{n=q} (\frac{x}{2})^n|$ so that

$$d_1(P_p, P_q) < |\sum_{n=p+1}^{n=q} (\frac{1}{2})^n| = \frac{1}{2^p}(1 - (\frac{1}{2})^{q-p}).$$

This shows that $\{P_N\}$ is a Cauchy sequence for d_1.

Alternatively, let us suppose that P_N converge towards an element $P \in \mathbb{R}[X], P(x) = \sum_{n \leq d} a_n x^n$. If we extend d_1 to the space of continuous functions $[0,1] \to \mathbb{R}$ and if we consider the function g defined on $\mathbb{R} \setminus \{2\}$ by $x \to g(x) = \frac{2}{2-x}$, we have:

$$
\begin{aligned}
d_1\left(P_N, \frac{2}{2-x}\right) &= \max_{x \in [0,1]} \left(P_N(x) - \frac{2}{2-x}\right), \\
&= \max_{x \in [0,1]} \left(P_N(x) - \sum_{n=0}^{\infty} \left(\frac{x}{2}\right)^n\right), \\
&= \max_{x \in [0,1]} \left(\sum_{n>N} \left(\frac{x}{2}\right)^n\right), \\
&= \sum_{n>N} \left(\frac{1}{2}\right)^n, \\
&= \frac{1}{2^{N+1}} \sum_{n \geq 0} \left(\frac{1}{2}\right)^n, \\
&= \frac{1}{2^N}.
\end{aligned}
$$

This means that P_N tends toward g when $N \to \infty$ for d_1.
This implies:

$$
P(x) = \sum_{n \leq d} a_n x^n = \frac{2}{2-x}.
$$

From the fundamental theorem of Algebra, $P(x)$ must have at least one root $z \in \mathbb{C}$, e.g. $\exists z \in \mathbb{C}, P(z) = 0$ but there is not $z \in \mathbb{C}$ such that $\frac{2}{2-z} = 0$ and therefore P cannot exist.

We conclude that $(\mathbb{R}[X], d_1)$ is not complete.

b) We have $d_2 \leq d_1$ so that if $(\mathbb{R}[X], d_2)$ is complete, this will imply that $(\mathbb{R}[X], d_1)$ is also complete, and this is clearly impossible from a).

So that $(\mathbb{R}[X], d_2)$ is not complete.

c) $\forall (p,q) \in \mathbb{N}^2, d_3(P_p, P_q) = d_1(P_p, P_q) = \frac{1}{2^p}(1 - (\frac{1}{2})^{q-p})$ so that $\{P_N\}$ is Cauchy for d_3.

If $\{P_N\}$ would converge towards $P \in \mathbb{R}[X] = \sum_{n \le d} a_n x^n$ for d_3, then, for $N > d$:

$$d_3(P_N, P) = \sum_{i \le d} |a_i - \frac{1}{2^i}| + \sum_{d < n \le N} \frac{1}{2^n} > \frac{1}{2^{d+1}}.$$

And, clearly, this is impossible so that $(\mathbb{R}[x], d_3)$ is also not complete.

Notes

[i]We could have also extended d_3 to the space of formal sums $f(x) = \sum_{i=0}^{\infty} a_i x^i$ and we see that $d_3(P_N, g) = d_1(P_N, g) = \frac{1}{2^N}$.

34 bis/828

Subject: Compacity of the completion of a metric space

Let X be a metric space that has a finite ε-net for every $\varepsilon > 0$. Show that the completion of X is compact.

Extra-Hint: *Consider the completion of X as the adherence of X in its own completion.*

SOLUTION:

discussion:

This exercise was not present in the original (e.g. Russian) version of TPFA ([?]). It may be viewed as an immediate application of the Hausdorff Criterion for precompacity (*[?] - CH 3, §2, Theorem 10*): *If X is a complete metric space and A a subset of X. In order for A to be precompact, it is necessary and sufficient that the set A has a finite ε-net for every $\varepsilon > 0$.*

Knowing this and from the fact that if $\widetilde{X}$ is the completion of X and $adh(X)$ is the adherence of X in $\widetilde{X}$, we will have $\widetilde{X} = \text{adh}(X)$, we only have to apply the Hausdorff criterion to X seen as a subspace of $\widetilde{X}$.

solution:

Let us suppose that X has an ε-net Λ_ε for every $\varepsilon > 0$. Then $\widetilde{X}$ shares this ε-net with X for, if $\widetilde{x} \in \widetilde{X}$, we can find $x \in X$ such that $d(\widetilde{x}, x) < \varepsilon/2$ (x belongs to the Cauchy sequence whose limit is $\widetilde{x}$) and given this element x, we can always find an element x_0 in $\Lambda_{\varepsilon/2}$ such that $d(x, x_0) < \varepsilon/2$ From this we deduce that $d(\widetilde{x}, x_0) < \varepsilon$ and that $\widetilde{X}$ has an ε-net.

In order to prove the compacity of $\widetilde{X}$, we can prove that from every sequence in $\widetilde{X}$, we can extract a Cauchy subsequence. As $\widetilde{X}$ is complete, all
Cauchy sequences will converge and thus we could extract from every sub-
sequence a converging subsequence which is a criterion for completeness.
So we consider an infinite sequence $\widetilde{x}_i$ in $\widetilde{X}$.

We start with an $\Lambda_{1/2}$ net. We consider, in $\widetilde{X}$, the set S_1 made with disjoint open balls of radius 1/2 and center $\widetilde{x}_i$. If this set is infinite then the set
$\bigsqcup_{B \in S_1} B$ cannot be included in a finite 1/2-net, which is clearly impossible.
So S_1 must be finite what means that at least one open ball — let us say $B^{(1)}$
— of radius 1/2 and center $x^{(1)}$ contains an infinity of elements from the sequence. We shift now the original sequence to this infinite subsequence and
we iterate the process considering now the set S_2 of disjoint open balls of
radius 1/4 and centers from the sequence. Again we see that there must
exist some $B^{(2)}$ with center $x^{(2)}$ from the sequence that contains an infinite
subsequence. Now we have a sequence of balls $B^{(n)}$ with center $x^{(n)}$ that
creates a subsequence from the original sequence $\{x_n\}$ and with decreasing
radius $1/2^n$. We would like to use the *shrinking ball* theorem (see Exercise
24) but we cannot because these balls are not necessarily nested in each
other.

We must therefore build a shrinking ball sequence: we consider for this
purpose the balls $B'^{(n)}$ of center $x^{(n)}$ and radius twice the radius of $B^{(n)}$ -
that is to say $1/2^{n-1}$.

This new sequence is a shrinking ball sequence (see Fig. 11).

Indeed, if we calculate $d(x^{(1)}, x)$ for $x \in B^{(n)}$:

$$d(x^{(1)}, x) \leq d(x^{(1)}, x^{(2)}) + d(x^{(2)}, x^{(3)}) + \ldots + d(x^{(n-2)}, x^{(n-1)}) + d(x^{(n-1)}, x).$$

That is to say:

$$d(x^{(1)}, x) \leq \frac{1}{2} + \frac{1}{4} + \ldots + \frac{1}{2^{n-1}} < 1.$$

So that $x \in B'^{(1)}$ and we can easily see that we can use the same reasoning for $B'^{(2)}$, etc...

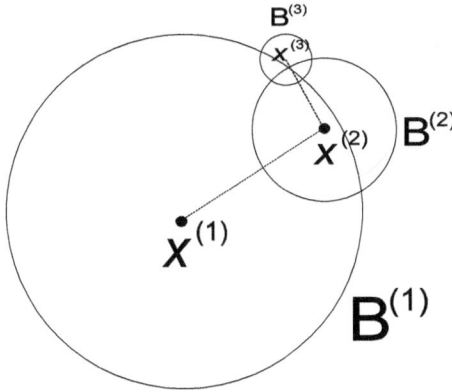

Figure 11: Set of balls $B^{(n)}$.

We can consider the set of closed balls $\overline{B'^{(n)}}$ - they are still shrinking - so that their intersection is a unique point x_{lim} that is the limit of the (sub)sequence $\{x^{(n)}\}$.

Notes

[i] The condition that the ε-net is finite for all the $\varepsilon > 0$ is mandatory.

If for some $\varepsilon > 0$ no finite ε-net can be found then we can define a sequence in the following way.

We take x_1 as any point in $\widetilde{X}$ and we consider a second point x_2 outside the open ball $B(x_1, \varepsilon)$. This point exist otherwise $B(x_1, \varepsilon)$ would be a finite ε-net. Again we build x_3 outside $B(x_1, \varepsilon) \cup B(x_2, \varepsilon)$. This point exist otherwise $B(x_1, \varepsilon) \cup B(x_2, \varepsilon)$ would be a finite ε-net, etc... At the end we have build an infinite sequence whose points are distant from each other at least ε. No Cauchy subsequence can then be extracted and thus $\widetilde{X}$ cannot be complete.

For example, $\mathbb{R}$ provided with the usual distance is a complete metric space but do not have any finite ε-net, then it is not compact.

35/828

Subject: Completeness of a functional space

On the space $\mathscr{C}(X,Y)$ of continuous applications from a metric space X onto a complete bounded metric space Y, we define a distance by:

$$d(f_1, f_2) = \sup_{x \in X} d_Y(f_1(x), f_2(x)).$$

Show that $\mathscr{C}(X,Y)$ is complete.

SOLUTION:

First we note that $d(f_1, f_2) < \infty$ since Y is bounded. From this, we can easily check that d is a distance over $\mathscr{C}(X,Y)$ from the fact that d_Y is a distance.

The demonstration is similar to the one in *Exercise 29*:

If $\{f_n\}$ is a Cauchy sequence in $\mathscr{C}(X,Y)$, then $\forall x, \{f_n(x)\}$ is also a Cauchy sequence in Y [55] and thus, tends toward a value $f(x)$ in Y because this space is complete: $f(x) = \lim_{n \to \infty} f_n(x)$.

The convergence $f_n \to f$ is uniform.

For an $\varepsilon > 0$, $\exists N_\varepsilon$ so that, $\forall N > N_\varepsilon$, we can find a subset $\{n_k\}$ with:

$$n_1 = N, \forall x, \forall k > 0, d_Y(f_{n_k}(x), f_{n_{k+1}}(x)) < d_{\mathscr{C}(X,Y)}(f_{n_k}, f_{n_{k+1}}) < \varepsilon/2^k.$$

[55] Indeed, $\forall x, \forall (p,q), d_Y(f_p(x), f_q(x)) \le d_{\mathscr{C}(X,Y)}(f_p, f_q)$

Indeed, we use the following method: f_n is a Cauchy sequence so that $\exists N_\varepsilon, (\forall (N_1, N_2) > N_\varepsilon), d(f_{N_1}, f_{N_2}) < \varepsilon/2$. We define n_1 by $n_1 = N$.

We have also that

$$\exists N_\varepsilon^{(1)}, (\forall (N_1^{(1)}, N_2^{(1)}) > N_\varepsilon^{(1)}), d(f_{N_1^{(1)}}, f_{N_2^{(1)}}) < \varepsilon/4$$

and then we define n_2 by $n_2 = N_1^{(1)}$, we iterate this process: for $k > 0$, $\exists N_\varepsilon^{(k)}, (\forall (N_1^{(k)}, N_2^{(k)}) > N_\varepsilon^{(k)}), d(f_{N_1^{(k)}}, f_{N_2^{(k)}}) < \varepsilon/2^{k+1}$ and we define n_{k+1} by $n_{k+1} = N_1^{(k)}$.

Then by summing all the k's, $k = 1, \ldots, \infty$ we will get:

$$\sum_{k=1}^{\infty} d_Y(f_{n_k}(x), f_{n_{k+1}}(x)) < \sum_{k=1}^{\infty} \frac{\varepsilon}{2^k}.$$

We also have $d(f_N(x), f_M(x)) \leq \sum_{k=1}^{k=M} d_Y(f_{n_k}(x), f_{n_{k+1}}(x))$ so that by making $M \to \infty$, we get:

$$d(f_N(x), f(x)) \leq \sum_{k=1}^{\infty} d_Y(f_{n_k}(x), f_{n_{k+1}}(x)) < \varepsilon.$$

So that:

$$\forall \varepsilon > 0, \exists N_\varepsilon, \forall N > N_\varepsilon, d_Y(f(x), f_N(x)) < \varepsilon.$$

Let us show that f is continuous: we have, $\forall N > 0, \forall (x, y) \in X^2$:

$$d_Y(f(x), f(y)) \leq d_Y(f(x), f_N(x)) + d_Y(f_N(x), f_N(y)) + d_Y(f_N(y), f(y))$$

$\forall \varepsilon > 0$, we can find N such that, $\forall n > N$, $\forall x \in X$ [56]:

$$d_Y(f(x), f_N(x)) < \varepsilon/3.$$

We can also find $\delta > 0$ [57] such that:

---

[56] Because of the uniform convergence $f_n \to f$

[57] Because f_N is continuous

$$d_X(x,y) < \delta \implies d_Y(f_N(x), f_N(y)) < \varepsilon/3.$$

Finally, $\forall \varepsilon > 0, \exists \delta > 0$ such that:

$$d_X(x,y) < \delta \implies d_Y(f(x), f(y)) < \varepsilon.$$

This shows that $f \in \mathscr{C}(X,Y)$.

To conclude, every Cauchy sequence in $\mathscr{C}(X,Y)$ converge towards an element of $\mathscr{C}(X,Y)$ so that $\mathscr{C}(X,Y)$ is complete.

36/828

Subject: Completeness of a space of homeomorphisms

Let X be a complete bounded metric space provided with a metric d_X and let G the set of applications that are bijective and bicontinuous from X into X (*homeomorphisms*[58] *of X*).

We define a distance over G by:

$$d_G(f_1, f_2) = \sup_{x \in X} [d_X(f_1(x), f_2(x)) + d_X(f_1^{-1}(x), f_2^{-1}(x))].$$

Study the completeness of the metric space (G, d_G).

SOLUTION:

Let $\{f_n\}$ be a Cauchy sequence in G, then $\{f_n^{-1}\}$ is also a Cauchy sequence: indeed, $d_G(f_p, f_q) = d_G(f_p^{-1}, f_q^{-1})$. From *Exercise 35*, we know that $\mathscr{C}(X,X)$, the space of continuous functions from X into X, is complete for the distance $d_{\mathscr{C}(X,X)}$ defined by $d_{\mathscr{C}(X,X)}(f,g) = \sup_{x \in X} d_X(f(x), g(x))$ and we see immediately that:

$$d_{\mathscr{C}(X,X)} \leq d_G$$

so that $\{f_n\}$ and $\{f_n^{-1}\}$ are Cauchy sequences in $\mathscr{C}(X,X)$ and therefore, since $\mathscr{C}(X,X)$ is complete, each of the two sequences must converge, for

---

[58] From Bourbaki, General Topology, Theorem 2, 1.10.2, *the homeomorphisms of a topological space X are the bicontinuous bijections of X*

the distance $d_{\mathscr{C}(X,X)}$, toward an element of this space: $f_n \to f$ and $f_n^{-1} \to g$, $(f,g) \in \mathscr{C}(X,X)^2$.

Let us prove that $g = f^{-1}$:

$$
\begin{aligned}
d_X(f \circ g(x), x) &= d_X(f \circ g(x) - f \circ f_n^{-1}(x) + f \circ f_n^{-1}(x), f_n \circ f_n^{-1}(x)), \\
&\leq d_X(f \circ g(x), f \circ f_n^{-1}(x)) + d_X(f \circ f_n^{-1}(x), f_n \circ f_n^{-1}(x)), \\
&\leq d_X(f \circ g(x), f \circ f_n^{-1}(x)) + d_{\mathscr{C}(X,X)}(f, f_n).
\end{aligned}
$$

We see that:

$$
d_X(f \circ g(x), f \circ f_n^{-1}(x)) \to d_X(f \circ g(x), f \circ g(x)) = 0 , \; n \to \infty \; ^{59}
$$

and that:

$$
d_{\mathscr{C}(X,X)}(f, f_n) \to d_{\mathscr{C}(X,X)}(f, f) = 0, \;, n \to \infty
$$

so that, by making $n \to \infty$:

$$
\forall x \in X, d_X(f \circ g(x), x) = 0.
$$

This proves that $f = g^{-1}$ and that $f \in G$ so that, to conclude, G is complete.

Since $d_G(f_1, f_2) \leq d_{\mathscr{C}(X,X)}(f_1, f_2) + d_{\mathscr{C}(X,X)}(f_1^{-1}, f_2^{-1})$, this implies that f_n converge towards f for d_G, and that f^{-1} is being defined, $f^{-1} \in \mathscr{C}(X,X)$. Then $f \in G$. This prove that (G, d_G) is complete.

---

^{59} Because f is continuous

37/828 𝓑

Subject: *p-adic* valuation

Let p be a prime number, we define a *p-adic* valuation over $\mathbb{Q}$ by[60]:

$$||0||_p = 0,$$

$$||r||_p = p^{-k}, r \neq 0, \text{ if } r = p^k \frac{m}{n}, \Delta(m,p) = 1 \text{ and } \Delta(n,p) = 1 \text{ [61]}.$$

Prove the following relations:

a) $||r_1 r_2||_p = ||r_1||_p ||r_2||_p$;

b) $||r_1 + r_2||_p \leq \max(||r_1||_p, ||r_2||_p)$: the *Ultrametric* inequality;

c) $||r_1||_p < ||r_2||_p \implies ||r_1 + r_2||_p = ||r_2||_p$.

SOLUTION:

If $r_1 = p^{k_1} \frac{m_1}{n_1}$ and $r_2 = p^{k_2} \frac{m_2}{n_2}$ with $\Delta(m_i, p) = 1$ and $\Delta(n_i, p) = 1, i \in \{1,2\}$ then:

---

[60]The precision that $||0||_p = 0$ is absent from TPFA

[61]$\Delta(a,b) = 1$ means that a and b are primes to each other

a)

$$r_1 r_2 = p^{k_1+k_2} \times \frac{m_1 \times m_2}{n_1 \times n_2} \text{ with } \Delta(m_1 \times m_2, p) = 1 \text{ and } \Delta(n_1 \times n_2, p) = 1.$$

So that:

$$||r_1 r_2||_p = p^{-(k_1+k_2)} = ||r_1||_p ||r_2||_p.$$

b)

If $k_1 = k_2$ then $r_1 + r_2 = p^{k_1} \frac{m_1 n_2 + m_2 n_1}{n_1 n_2}$. $\Delta(n_1, p) = 1$ and $\Delta(n_2, p) = 1$ so that $\Delta(n_1 n_2, p) = 1$. Let us suppose that $m_1 n_2 + m_2 n_1 = p^\alpha k$, with $\Delta(k, p) = 1$, then $r_1 + r_2 = p^{-k_1+\alpha} \frac{k}{n_1 n_2}$ and then: $||r_1 + r_2||_p = p^{-k_1-\alpha} \leq p^{-k_1} \leq \max(||r_1||_p, ||r_2||_p)$.

Let us suppose that $k_1 < k_2$:

$$
\begin{aligned}
r_1 + r_2 &= p^{k_1} \frac{m_1}{n_1} + p^{k_2} \frac{m_2}{n_2}, \\
&= p^{k_1} \left(\frac{m_1}{n_1} + p^{k_2-k_1} \frac{m_2}{n_2} \right), \\
&= p^{k_1} \left(\frac{m_1 n_2 + p^{k_2-k_1} m_2 n_1}{n_1 n_2} \right),
\end{aligned}
$$

$m_1 n_2 + p^{k_2-k_1} m_2 n_1 \equiv m_1 n_2 \neq 0 \pmod{p}$ so that $\Delta(m_1 n_2 + p^{k_2-k_1} m_2 n_1), p) = 1$. Since $\Delta(n_1 n_2, p) = 1$, we have:

$$||r_1 + r_2||_p = p^{-k_1} = \max(||r_1||_p, ||r_2||_p).$$

c)

In fact we proved in b) that, if $r_1 \neq r_2$, the p-adic valuation in $\mathbb{Q}$ satisfies:

$$||r_1 + r_2||_p = \max(||r_1||_p, ||r_2||_p).$$

38/828

Subject: The Field of p-adic $\mathbb{Q}_p$

Show that $\mathbb{Q}$ provided with the distance $d_p(r_1, r_2) = ||r_1 - r_2||_p$ is a metric space. Let $\mathbb{Q}_p$ be its completion. Show that the arithmetic operations in $\mathbb{Q}$ extends themselves continuously to $\mathbb{Q}_p$. $\mathbb{Q}_p$ is named the *p-adic field*.

SOLUTION:

First, we must check that $(\mathbb{Q}, d_p)$ is a metric space. Using the properties of the p-adic valuation $||.||_p$ (see Exercise 37), we get:

- $(\forall x \in \mathbb{Q}), d_p(x, x) = ||0||_p = 0$,
- $(\forall (x, y) \in \mathbb{Q}^2)$, if $x \neq y$ then: $d_p(x, y) = 0 \Rightarrow ||x - y||_p = 0 \Rightarrow x = y$ [62],
- $(\forall (x, y, z) \in \mathbb{Q}^3), d_p(x, y) + d_p(y, z) = ||x - y||_p + ||y - z||_p \geq ||x - y + y - z||_p = d_p(x, z)$.

This proves that d_p is a metric for $\mathbb{Q}$.

We now consider the sequences $\{x_n\}$ and $\{y_n\}$ that we suppose to be Cauchy for d_p.

We have for $(a, b) \in \mathbb{N}^2$:

$$\begin{aligned}
d_p(x_a + y_a, x_b + y_b) &= ||x_a + y_a - x_b - y_b||_p \\
&\leq \max(||x_a - x_b||_p, ||y_a - y_b||_p)
\end{aligned}$$

and:

---

[62] We see that $(\forall z \in \mathbb{Q}), ||z||_p = 0 \Rightarrow z = 0$. Indeed, if $z \neq 0$ then $||z||_p = p^{-k}, k \in \mathbb{Z}$ and $||z||_p \neq 0$

$$
\begin{aligned}
d_p(x_a y_a, x_b y_b) &= ||x_a y_a - x_b y_b||_p, \\
&= ||x_a y_a - x_a y_b + x_a y_b - x_b y_b||_p, \\
&\leq ||x_a y_a - x_a y_b||_p + ||x_a y_b - x_b y_b||_p, \\
&\leq ||x_a(y_a - y_b)||_p + ||y_b(x_a - x_b)||_p, \\
&\leq ||x_a||_p ||y_a - y_b||_p + ||y_b||_p ||x_a - x_b||_p, \\
&\leq (\sup_{n \geq 0}||x_n||_p)||y_a - y_b||_p + (\sup_{n \geq 0}||y_n||_p)||x_a - x_b||_p,
\end{aligned}
$$

$(\sup_{n \geq 0}||y_n||_p$ is well-defined: if x_n is a Cauchy sequence for d_p, then $\sup_{n \geq 0}||x_n||_p) < \infty$ otherwise x_n would not converge for d_p: if $\forall M > 0, \forall N > 0, \exists n \geq N, ||x_n||_p > M$ then we could build a sequence x_{n_k} such that $x_{n_{k+1}} > x_{n_k} + 1$, which is impossible).

This shows that $\{x_n + y_n\}$ and $\{x_y y_n\}$ are also Cauchy sequences so that we can extend continuously the operations "+" and "×" from $\mathbb{Q}$ to $\mathbb{Q}_p$.

Let us suppose that $x_n \not\to 0$, that is to say: $\exists N$ such that $\forall n > N \ ||x_n||_p \geq c > 0$:

$$
||\frac{1}{x_a} - \frac{1}{x_b}|| = \frac{||x_a - x_b||_p}{||x_a||_p ||x_b||_p} < \frac{||x_a - x_b||_p}{c^2}.
$$

This shows that $\{x_n^{-1}\}$ is also a Cauchy sequence so that the operation $x \to x^{-1}$ can be also continuously extended from $\mathbb{Q}$ to $\mathbb{Q}_p$.

Then the properties of a field are extended to $\mathbb{Q}_p$ so that this is a field too.

39/828 ¶

Subject: representation of p-adic elements as formal sums

Show that any element of the p-adic field $\mathbb{Q}_p$ can be represented in an unique way by a p-adic fraction:

$$\ldots a_2 a_2 a_0, a_{-1} a_{-2} \ldots a_{-k}.$$

Where $0 \leq a_i \leq p-1$ and there are an infinite number of a_i's.

In other words, every element in $\mathbb{Q}_p$ is the sum of a convergent[63] sum: $\sum_{i=-k}^{+\infty} a_i p^i$.

Show that $\mathbb{Q}_p \neq \mathbb{Q}$.

SOLUTION:

The sequences $\{S_N\} = \{\sum_{i \leq N} a_i p^i\}$ are Cauchy sequences because:

$$d_p(S_a, S_b) = || \sum_{a < i \leq b} a_i p^i ||_p \leq \max_{a < i \leq b} ||a_i p^i||_p = p^{-a} \ ^{64}.$$

So that $S_N \to S = \sum_{i=-k}^{+\infty} a_i p^i \in \mathbb{Q}_p \ (k \in \mathbb{Z}) \ ^{65}$.

---

[63] Convergent for d_p, not for the usual distance in $\mathbb{R}$

[64] In Exercise 37, we have proved that the p-adic valuation in $\mathbb{Q}$ had the following property: $||x+y||_p \leq \max(||x||_p, ||y||_p)$, this property can be obviously extended to $\mathbb{Q}_p$ by the completion process and we can also extend it to $x_1, \ldots, x_n \in \mathbb{Q}_p$: $||\sum_{i=1,\ldots,n} x_i||_p \leq \max_{i=1,\ldots,n}(||x_i||_p)$

[65] A criterion for a sum $\sum_{n \leq N} x_n$ to converge in $\mathbb{Q}_p$ is that $||x_n||_p \to 0$. Here we see that, obviously, $||a_N p^N||_p \to 0$

Let us now consider $x = \sum_{i=-k}^{+\infty} a_i p^i$ and $y = \sum_{i=-l}^{+\infty} b_i p^i$.
We have:

$$d(x,y) = \left|\left|\sum_{i=-\max(k,l)}^{+\infty} (a_i - b_i)p^i\right|\right|_p$$

(we might have to "pad" the a_i's or the b_i's with zeros so that they both reach the rank $\max(k,l)$).

If $i_0 = \min\{i, a_i \neq b_i\}$ then:

$$||(a_{i_0} - b_{i_0})p^{i_0}||_p = \max_{i=-\max(k,l)\ \dots\ +\infty} ||(a_i - b_i)p^i||_p.$$

So that:

$$d(x,y) = p^{-i_0}.$$

This shows the uniqueness of the representation by such sums.
We consider the set $\mathbf{Q}'$ defined by:

$$\mathbf{Q}' = \{x \in \mathbb{Q}_p, x = \sum_{i=-k}^{+\infty} a_i p^i, 0 \leq a_i \leq p-1\}.$$

If $\{x_n\}$ is a Cauchy sequence in $\mathbf{Q}'$, we can write:

$$x_n = \sum_{i=-k_n}^{+\infty} a_i^{(n)} p^i.$$

In that case, we can also write:

$$d(x_s, x_t) = p^{-\min\{i, a_i^{(s)} \neq a_i^{(t)}\}}.$$

So that, if $(s,t) \to \infty$, we must have $\min\{i, a_i^{(s)} \neq a_i^{(t)}\} \to \infty$.
This means that:

$$\forall i, \exists N, n > N \implies a_i^{(n)} = a_i^{(N)} = \alpha_i.$$

In other words, after a given rank, the sequence $\{(a_i^{(n)})_n\}$ becomes constant.

We define n_1 so that, $\forall n > n_1, a_1^{(n)} = \alpha_1$ then we define $n_2 > n_1$ such that: $\forall n > n_2, a_2^{(n)} = \alpha_2$. We define similarly $n_3, \ldots, n_k$, etc...

If $x = \sum_N \alpha_N p^N$, then: $d_p(x_{n_k}, x) \le p^{-n_k}$ so that any Cauchy sequence in $\mathbf{Q}'$ converge in $\mathbf{Q}'$ and thus, this is a complete space.

Let us prove that $\mathbb{Q} \subset \mathbf{Q}'$.

If $x = \frac{m}{n}$, $(m, n) \in \mathbb{N}^2$ and $\Delta(n, p) = 1$ then we build a sequence $\{m_n, a_n\}$ defined by:

$$m_0 = m,$$

$$m_i - a_i n = m_{i+1} p, \ 0 \le a_i \le p - 1, \forall i \ge 0.$$

Indeed, this means that[66] :

$$
\begin{aligned}
\frac{m}{n} &= a_0 + p\frac{m_1}{n}, \\
\frac{m_1}{n} &= a_1 + p\frac{m_2}{n}, \\
&\cdots \\
\frac{m_N}{n} &= a_N + p\frac{m_{N+1}}{n}, \\
&\cdots
\end{aligned}
$$

so that

$$x = \frac{m}{n} = \sum_N a_N p^N.$$

This shows that $\mathbb{Q} \subset \mathbf{Q}' \subset \mathbb{Q}_p$ and we also know that $\mathbf{Q}'$ is complete so that, to conclude: $\mathbf{Q}' = \mathbb{Q}_p$.

The field of the p-adic is the field of the formal sums $\sum_N a_N p^N$.

Let us show now that $\mathbb{Q} \ne \mathbb{Q}_p$.

---

[66]Following Bezout's theorem: *if α and β are such that $\Delta(\alpha, \beta) = 1$, then $\forall N \in \mathbb{Z}$, we can find $A \in \mathbb{Z}$ and $B \in \mathbb{Z}$ such that $A\alpha + B\beta = N$ (here $\alpha = n, \beta = p, A = a_i, B = m_{i+1}, N = m_i$)*

First, Let us suppose that for a $x \in \mathbb{Q}_p$, after a rank k, we have a period of length N, that is to say:

$$x = a_0 \ldots a_{k-1} \boxed{a_k a_{k+1} \ldots a_{k+N-1}} \boxed{a_k a_{k+1} \ldots a_{k+N-1}} \boxed{a_k a_{k+1} \ldots a_{k+N-1}} \ldots$$

then we have:

$$x = (a_0 + a_1 p \ldots + a_{k-1} p^{k-1}) + (a_k p^k + \ldots + a_{k+N-1} p^{k+N-1})(1 + p^N + p^{2N} + \ldots).$$

In $\mathbb{Q}_p$, we have:

$$\sum_{\alpha=0}^{\alpha=\infty} p^{\alpha N} = \frac{1}{1 - p^N}.$$

Indeed, $||\sum_{\alpha \leq \alpha_0} p^{\alpha N} - \frac{1}{1-p^N}||_p = ||\sum_{\alpha > \alpha_0} p^{\alpha N}||_p = p^{-N(\alpha+1)}$.
So that:

$$x = (a_0 + a_1 p \ldots + a_{k-1} p^{k-1}) + p^k(a_k + a_{k+1} p + \ldots + a_{k+N-1} p^{N-1})(\frac{1}{1 - p^N}).$$

This shows that $x \in \mathbb{Q}$.

Conversely, let us show that, for $x = \frac{m}{n}$, $\Delta(n,p) = 1$ we can find U, V, N and k so that:

$$U = a_0 + a_1 p \ldots + a_{k-1} p^{k-1},$$
$$V = a_k + a_{k+1} p + \ldots + a_{k+N-1} p^{N-1},$$
$$\frac{m}{n} = U + p^{k-1} V (\frac{1}{1 - p^N}).$$

If $m' = m - nU$, then this is equivalent to:

$$m' \times (1 - p^N) = n \times p^{k-1} V.$$

From Euler's theorem, we know this, because $\Delta(n,p) = 1$:

$$1 - p^{\varphi(n)} \equiv 0 \ (n).$$

We may define U and m' to be such that $m = m' + nU$ (Euclidean division of m by n).

We put $N = \varphi(n)$ and then, by Euler's theorem, $1 - p^N = 1 - p^{\varphi(n)} = rn, r \in \mathbb{Z}$.

From here, we define V and k to be such that $m'r = p^{k-1}V$. That is to say: $p^{-k+1} = ||m'r||_p, r \in \mathbb{Z}, \Delta(p,V) = 1$. Then because $m'r \in \mathbb{Z}$, we will have $k - 1 \geq 0$.

In $\mathbb{Z}_p$, we have $x = U + \frac{m'}{n} = U + p^{k-1}\frac{V}{rn} = U + p^{k-1}\frac{V}{1-p^N} = U\boxed{V}\boxed{V}\ldots\boxed{V}\ldots$

The period might be smaller than $\varphi(n)$ but in any case the development is $\varphi(n)$-periodic, this shows, that p-adic developments that are periodic after a certain rank represent the rational numbers so that the non-periodic developments are not in $\mathbb{Q}$ and, therefore $\mathbb{Q} \neq \mathbb{Q}_p$.

40/828

Subject: Properties of $\mathbb{Q}_5$

Show that, in the field of the 5-adic $\mathbb{Q}_5$, we have the following properties:

$2+3 = \ldots \boxed{0}\,0010$;

$2-3 = \ldots \boxed{4}\,4444$;

$2 \times 3 = \ldots \boxed{0}\,0011$;

$2/3 = \ldots \boxed{31}\,31314$.

SOLUTION:

1) $2 = 2 + \sum_{n \geq 1} 0 \times 5^n = \ldots 000002$ and $3 = 3 + \sum_{n \geq 1} 0 \times 5^n = \ldots 000003$
so that $2+3 = 5 = 0 + 1 \times 5 + \sum_{n \geq 2} 0 \times 5^n = \ldots 000010$.

2) $3 - 2 + \ldots \boxed{4}\,4444 = 1 + \ldots \boxed{4}\,4444 = \ldots \boxed{0}\,0000 = 0$.

3) $2 \times 3 = (2 + \sum_{n \geq 1} 0 \times 5^n) \times (3 + \sum_{n \geq 1} 0 \times 5^n) = 1 + 1 \times 5 + \sum_{n \geq 2} 0 \times 5^n = \ldots \boxed{0}\,0011$.

4) we build the sequence m_i, a_i described in *Exercise 39*:

$m_0 = 2$

$$
\begin{aligned}
m_0 &= a_0 n + m_1 p, \\
2 &= 3a_0 + 5m_1, \\
a_0 &= 4, m_1 = -2,
\end{aligned}
$$

$$m_1 = a_1 n + m_2 p,$$
$$-2 = 3a_1 + 5m_2,$$
$$a_1 = 1, m_2 = -1,$$

$$m_2 = a_2 n + m_3 p,$$
$$-1 = 3a_2 + 5m_3,$$
$$a_2 = 3, m_3 = -2.$$

We have

$$m_3 = m_1$$

so that we will have a 2-length period after the rank 1:

$$2/3 = \dots \boxed{31}\,31314.$$

We check this by a direct calculation:

$$\begin{aligned}
\dots \boxed{31}\,31314 &= 4 + (1 \times 5 + 3 \times 5^2)(1 + 5^2 + 5 + 2^4 + \dots), \\
&= 4 + \frac{5 + 3 \times 25}{1 - 5^2}, \\
&= 4 - \frac{80}{24}, \\
&= \frac{16}{24}, \\
&= \frac{2}{3}.
\end{aligned}$$

41/828

Subject: Square root of -1 in $\mathbb{Q}_5$

Show that, in $\mathbb{Q}_5$, the field of 5-adic numbers, we can extract the square root of $-1 = \ldots \boxed{4} 4444$.

SOLUTION:

We try to find $\sqrt{-1}|_{\mathbb{Q}_5} = \sum_{i \geq 0} a_i 5^i, 0 \leq a_i \leq 4$.
    We must have:

$$\left(\sum_{i \geq 0} a_i 5^i\right) \times \left(\sum_{i \geq 0} a_i 5^i\right) = -1$$

that is to say:

$$a_0^2 + (2a_0 a_1)5 + (a_1^2 + 2a_2 a_0)5^2 + \ldots = -1.$$

This gives the following conditions:

$$a_0^2 \equiv -1 \ (5),$$
$$a_0^2 + (2a_0 a_1)5 \equiv -1 \ (5^2),$$
$$a_0^2 + (2a_0 a_1)5 + (a_1^2 + 2a_2 a_0)5^2 \equiv -1 \ (5^3),$$
$$\ldots$$

or, for the first three coefficients $(a_0, a_1, a_2) \in [0, 4]$:

$$a_0^2 \equiv -1\ (5),$$
$$a_0^2 + 10a_0a_1 \equiv -1\ (25),$$
$$a_0^2 + 10a_0a_1 + 25a_1^2 + 50a_2a_0 \equiv -1\ (125),$$

or:

$$a_0^2 \equiv -1\ (5),$$
$$a_0^2 + 10a_0a_1 \equiv -1\ (25),$$
$$(a_0 + 5a_1)^2 + 50a_0a_2 \equiv -1\ (125).$$

We compute the values of a_0^2 in $\mathbb{Z}/5\mathbb{Z}$:

a_0	0	1	2	3	4
a_0^2	0	1	4	4	1

so that $a_0 = 2$ or $a_0 = 3$.
1) We consider the case $a_0 = 2$.
We must solve the following equivalent equations for finding a_1:

$$a_0^2 + 10a_0a_1 \equiv -1\ (25),$$
$$4 + 20a_1 \equiv -1\ (25),$$
$$20a_1 \equiv -5\ (25).$$

We compute the values of $20a_1$ in $\mathbb{Z}/25\mathbb{Z}$:

a_1	0	1	2	3	4
$20a_1$	0	20	15	10	5

there is a unique solution $a_1 = 1$.
We must solve the following equivalent equations for finding a_2:

$$(a_0 + 5a_1)^2 + 50a_0a_2 \equiv -1 \ (125),$$
$$(2+5)^2 + 100a_2 \equiv -1 \ (125),$$
$$100a_2 \equiv -50 \ (125).$$

We compute the values of $100a_2$ in $\mathbb{Z}/125\mathbb{Z}$:

a_2	0	1	2	3	4
$100a_2$	0	100	75	50	25

there is a unique solution $a_1 = 2$.
Let us show that the a_i's are solutions of:

$$(a_0 + 5a_1 + \ldots 5^{k-2}a_{k-2})^2 + 2 \times 5^{k-1}a_0a_{k-1} \equiv -1(5^k), \forall k > 1.$$

We know that

$$(\sum_{i \geq 0} a_i 5^i)^2 = -1$$

or:

$$\sum_{s=0}^{s=+\infty} (\sum_{i+j=s} a_i a_j) 5^s = -1$$

so that we have the relations:

$$\sum_{s=0}^{s=k-1} (\sum_{i+j=s} a_i a_j) 5^s \equiv -1(5^k), k > 1.$$

On the other hand, we have that:

$$(a_0 + 5a_1 + \ldots 5^{k-2}a_{k-2})^2 \;=\; \sum_{0 \le (i,j) \le k-2} a_i a_j 5^{i+j},$$

$$=\; \sum_{s=0}^{s=2(k-2)} \left(\sum_{i+j=s,(i,j) \le k-2} a_i a_j \right) 5^s,$$

$$\equiv\; \sum_{s=0}^{s=k-1} \left(\sum_{i+j=s,(i,j) \le k-2} a_i a_j \right) 5^s \;(mod\; 5^k),$$

$$\equiv\; \sum_{s=0}^{s=k-1} \left(\sum_{i+j=s} a_i a_j \right) 5^s - 2a_0 a_{k-1} 5^{k-1} \;(mod\; 5^k),$$

$$\equiv\; -1 - 2a_0 a_{k-1} 5^{k-1} \;(mod\; 5^k).$$

What proves the assertion.
Let us now show that the system

$$(a_0 + 5a_1 + \ldots 5^{k-2}a_{k-2})^2 + 2 \times 5^{k-1} a_0 a_{k-1} \equiv -1(5^k), \forall k > 1$$

has a unique solution $\{a_k\}_{k \in \mathbb{N}}$, a_0 and a_1 being given.
This system is equivalent to[67]:

$$(a_0 + 5a_1 + \ldots + 5^{k-3}a_{k-3})^2 + (5^{k-2}a_{k-2})^2$$
$$+ 2 \times (a_0 + 5a_1 + \ldots + 5^{k-3}a_{k-3})(5^{k-2}a_{k-2}) + 2 \times 5^{k-1} a_0 a_{k-1} \equiv -1(5^k), \forall k > 1$$

$\Leftrightarrow$

$$-1 + 5^{k-1}S_{k-1} - 2 \times 5^{k-2} a_0 a_{k-2}$$
$$+ 2 \times (a_0)(5^{k-2}a_{k-2}) + 2 \times 5^{k-1} a_0 a_{k-1} \equiv -1(5^k), \forall k > 1$$

$\Leftrightarrow$

$$-1 + 5^{k-1}S_{k-1} + 2 \times 5^{k-1} a_0 a_{k-1} \equiv -1(5^k), \forall k > 1.$$

---

[67] $S_{k-1} = 2(a_1 + \cdots + 5^{k-2}a_{k-3})a_{k-2}$

This leads to solve the linear equations for $k = 1, \ldots$ with $X = a_{k-1}$:

$$5^{k-1} S_{k-1} + 2 \times 5^{k-1} a_0 X \equiv 0(5^k).$$

For every $k \geq 1$, if we put $\alpha = 2 \times a_0$ and $\beta = S_{k-1}$, a_{k-1} is solution of the following linear equation:

$$\alpha X + \beta = 0, X \in \mathbb{Z}/5\mathbb{Z}.$$

This solution is unique[68] and such that:

$$X = -\frac{\beta}{\alpha}.$$

So that, by solving recursively the linear equations for all values of k, we get a unique solution $\{a_k\}_{k\in\mathbb{N}}$:

$$\sqrt{-1}|_{\mathbb{Q}_5} = \ldots 212.$$

2) the same process allows us to find a unique square root of -1 when $a_0 = 3$.

---

[68] Because $\mathbb{Z}/5\mathbb{Z}$ is a field and $\alpha \neq 0$

42/828 ¶

Subject: Homeomorphism between $\mathbb{Z}_p$ and the Triadic Cantor Set

We note $\mathbb{Z}_p$ the closure of the ring $\mathbb{Z}$ into $\mathbb{Q}_p$. Show that $\mathbb{Z}_p$ is a compact set. Build a bicontinuous bijection from $\mathbb{Z}_p$ into the triadic Cantor set C [69]

.

SOLUTION:

It should be rather clear from its definition as the closure of $\mathbb{Z}$ in $\mathbb{Q}_p$ that $\mathbb{Z}_p$ is represented by the set $\{\sum_{i \geq 0} a_i p^i, a_i \in [|0, p-1|]\}$.

In order to prove the compactness of $\mathbb{Z}_p$, we use the Hausdorff criterion that stipulates that, *if X is a complete metric space and A a subset of X. In order for A to be precompact, it is necessary and sufficient that the set A have a finite ε-net for every $\varepsilon > 0$.*

We consider the set:

$$\Sigma_N = \{\sum_{i=0}^{i=N} a_i p^i, 0 \leq a_i \leq p-1\}.$$

Then, for every $x \in \mathbb{Z}_p$, we can find $s_N \in \Sigma_N$ such that:

$$||x - s_N||_p \leq p^{-(N+1)}.$$

---

[69] We will also note this set C_2 in what follows

We have[70] $\#\Sigma_N = p^N$ so that Σ_N is a finite ε-network of $\mathbb{Z}_p$, where $\varepsilon > p^{-(N+1)}$.

This shows that $\mathbb{Z}_p$ has its closure compact[71] and, by definition, it is a closed set so that finally, $\mathbb{Z}_p$ is compact.

Regarding now the building of the homeomorphisms between $\mathbb{Z}_p$ and the Ternary/Triadic Cantor Set C, we recall that C is build in the following way (see fig 12):

We define two numeric functions f_1 and f_2 by:

$$f_1(x) = x/3, f_2(x) = (x+2)/3$$

and F, a function from the set of non-empty compact real subsets $K(\mathbb{R})$ into itself by:

$$F(X) = f_1(X) \cup f_2(X).$$

Then C is defined as the unique solution of:

$$F(X) = X, X \in K(\mathbb{R}).$$

That is to say that C is the limit of the the sequence[72]:

$$C = \lim_{n \to \infty} F^{(n)}([0,1]) = \bigcap_n F^{(n)}([0,1]).$$

Case $\mathbb{Z}_2$.

First we remark that $[0,1]$ may be described by the elements $\{x = \frac{1}{3}\sum_{i \geq 0} a_i 3^{-i}, a_i \in [|0,2|]\}$.

There exists an homeomorphism between $\mathbb{Z}_2$ and C that can be build as follows.

The zone defined by $f_1([0,1]) \cup f_2([0,1])$ can be described by:

$$\{x \in [0,1], x = \frac{1}{3}\sum_{i \geq 0} a_i 3^{-i}, a_0 = 0 \text{ or } a_0 = 2\}.$$

---

[70]We will note $\#A$ the number of element of element of a set A

[71]E.g. $\mathbb{Z}_p$ is pre-compact

[72]$F^{(n)}$ being the nth iterate of F, e.g. $F \circ \ldots \circ F$ n times

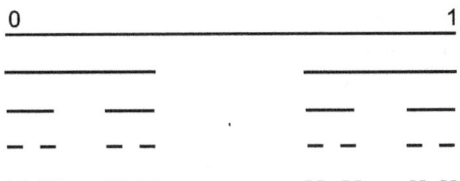

Figure 12: Building of the Ternary cantor set (the limit of remaining intervals being the Cantor Set)

So that $a_0 = 0$ implies that $x \in f_1([0,1])$ and $a_0 = 1$ implies that $x \in f_2([0,1])$.

Recursively, we can apply the same reasoning to $F^{(n)}([0,1])$:

$$F^{(n)}([0,1]) = \{x \in [0,1], x = \frac{1}{3}\sum_{i \geq 0} a_i 3^{-i}, \; a_j \in \{0,2\} \; \forall j = 0,\ldots,n-1\}.$$

Then, the Ternary Cantor Set C can be described by:

$$C = \{x \in [0,1], x = \frac{1}{3}\sum_{i \geq 0} a_i 3^{-i}, \; a_j \in \{0,2\} \; \forall j \geq 0\}.$$

We define the application φ from $\mathbb{Z}_2$ into C by:

$$\varphi(\sum_{N \geq 0} a_N 2^N) = \frac{1}{3}\sum_{N \geq 0} 2a_N 3^{-N}.$$

Since elements in $\mathbb{Z}_2$ have a unique decomposition $x = \sum_{N \geq 0} A_N 2^N$, φ is well-defined as an application and since any object in C has also a unique decomposition $x = \sum_{N \geq 0} A'_N 3^{-N}$, φ is an injection. We may define $\varphi^{-1} : C \to \mathbb{Z}_2$ by

$$\varphi^{-1}(\frac{1}{3}\sum_{N \geq 0} a_N 3^{-N}) = \sum_{N \geq 0} \frac{a_N}{2} 2^N$$

so that φ is a bijection.

Let us now prove that φ is bicontinuous.

If $(x,y) \in (\mathbb{Z}_2)^2$ and $||x-y||_2 < \varepsilon$, then we may write for N such that $\varepsilon > 2^{-N}$:

$$x = \alpha + \sum_{n \geq N} a_n 2^n$$

and

$$y = \alpha + \sum_{n \geq N} b_n 2^n.$$

What implies:

$$|\varphi(x) - \varphi(y)| = \frac{1}{3} | \sum_{n \geq N} 2(a_n - b_n)3^{-n}| < \frac{1}{3} \sum_{n \geq N} 2 \times 3^{-n} < 2 \frac{1}{3} \frac{1}{1 - \frac{1}{3}} 3^{-N} = 3^{-N}.$$

For $\delta > 0$, if N is being defined such that $3^{-N} < \delta$ and if $\varepsilon > 0$ is such that $\varepsilon > 2^{-N}$, then:

$$||x-y||_2 < \varepsilon \Rightarrow |\varphi(x) - \varphi(y)| < \delta.$$

This proves that φ is continuous. The proof that φ^{-1} is also continuous is very similar.

We conclude that φ is bicontinuous $\mathbb{Z}_2 \to C : \mathbb{Z}_2 \approx C$.

In a very similar way, we could build an homeomorphism: $\mathbb{Q}_2 \approx C - \{1\}$.

General case.

Now we need to build the homeomorphism in the general case, between C and $\mathbb{Z}_p$, and this looks much more difficult[73].

---

[73]This is equivalent, of course, to building an homeomorphism between $\mathbb{Z}_2$ and $\mathbb{Z}_p$

First try.

We see that every number in the range from 0 to $p-1$ can be represented by a finite binary sequence corresponding to its unique base 2 decomposition. Thus any p-sequence[74] of numbers ranging from 0 to $p-1$ can be mapped to a sequence of binary numbers:

e.g. $a_n \in [|0, p-1|], b_{n,i} \in \{0, 1\}, i = 1, \ldots k_n$ such that $a_n = \sum_{i=0}^{i=k_n} b_{n,i} 2^i$,

then we define the following association from $[|0, p-1|]$ into $\bigcup_{i=0}^{i=N_p} 2^i$:

$$a_n \to (b_{n,1}, \ldots b_{n,k_n}).$$

From this, we can deduce an application from p^∞ into 2^∞ as such:

$$\{a_n\}_{n \geq 0} \to (b_{1,1}, \ldots b_{n,k_1} \ldots b_{n,1}, \ldots b_{n,k_n} \ldots).$$

We have defined a mapping φ from p^∞ to 2^∞. This is a continuous surjective application but it is not a bijection: obviously two different p-sequences can produce the same binary sequence. If we try similar ways to deduce a binary sequence from a p-sequence (or the opposite) we see that unfortunately, in all cases, the two sets are not proportional and so it does not work.

Second try.

Changing our strategy, we consider now that we could simply establish a one-to-one correspondence between $\mathbb{Z}_2$ and $\mathbb{Z}_p$ simply by making "formally":

$$\sum_{n=0}^{n=\infty} a_n 2^n = \sum_{n=0}^{n=\infty} b_n p^n.$$

The problem is that we need to define b_0 as $\sum_{n=0}^{n=\infty} a_n 2^n \ (p)$. We cannot define it. If we consider the partial sums $S_N = \sum_{n=0}^{n=N} a_n 2^n \ (p)$ but we see that S_N will take an infinite number of times the value k, when $N \to \infty$,

---

[74]We call a *p*-sequence an element of $p^\infty = [|0, p-1|]^\infty$ and equivalently the decomposition of real numbers in base *p*

where $0 \leq k \leq p-1$.

Third try.

The only way to make the previous idea right would be to use $[0,1]$ as an intermediary ground for both $\mathbb{Z}_2$ and $\mathbb{Z}_p$ and we could use the fact that the a real number have both a decomposition as a binary sequence and as a p-sequence.

There is a well-known application from C to $[0,1]$ — or equivalently from $\mathbb{Z}_2$ to $[0,1]$ — called the Cantor's ladder and defined by:

$$\mathbb{Z}_2 \rightarrow [0,1],$$

$$\sum_{i=0}^{\infty} a_i 2^i \rightarrow \frac{1}{2} \sum_{i=0}^{\infty} a_i 2^{-i} , a_i \in \{0,1\}.$$

This application is not a bijection because some real numbers have two equivalent expression as binary sequences. Namely only the rational numbers of the form $\frac{r}{2^s}$ where r and s are integers, can be expressed as both different binary sequences. This comes from the fact that we have the equality $1 = 0.111111\ldots$ In the same spirit, every number of the form $\frac{r}{p^s}$ will be expressed by two distinct p-sequences because $1 = 0.(p-1)(p-1)(p-1)(p-1)\ldots$

So we cannot build directly an application from $\mathbb{Z}_2$ into $[0,1]$ then from $[0,1]$ into $\mathbb{Z}_p$. We should consider, prior to identify a binary sequence and a p-sequence, to split numbers of the form $\frac{r}{2^s}$ into both distinct binary sequences.

If we still want to build our application φ, we shall make a distinction between binary sequences that are infinitely "padded" with 1's and the others - and so we shall map them differently. This will be completely discontinuous from the point of view of $[0,1]$ but not necessarily from the p-adic point of view.

To explain our choice, we consider the topological operation $\mathbf{T}(x)$ of choosing a point x in $[0,1]$ and to use it to split $[0,1]$ in two closed parts (see Fig. 13). We can see the Cantor Set as the result of applying $\mathbf{T}(x)$

an infinite number of times to the points $x = \frac{r}{2^s}$ in $[0,1]$. Therefore, two equivalent 2-adic representation in [0,1] will be separated in two disjoint entities.

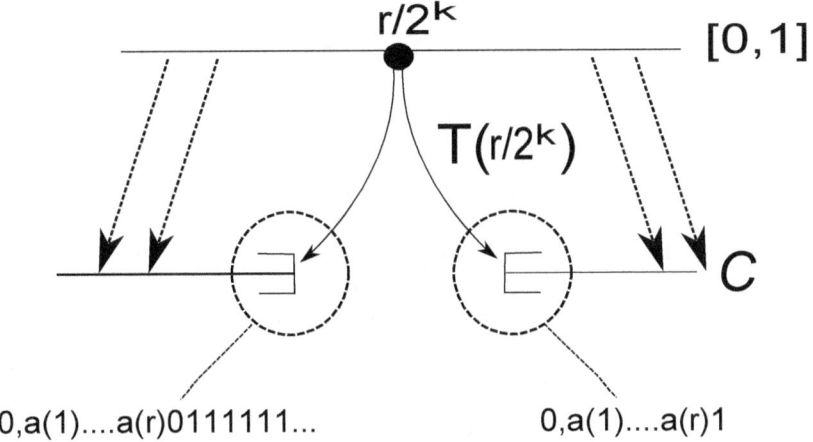

0,a(1)....a(r)0111111...   0,a(1)....a(r)1

Figure 13: Creation of the Cantor set by topological operation T(x)

We consider the "generalized" Cantor Set C_p defined for $p > 1$ by the same process than C except than one divide $[0,1]$ in $2p-1$ intervals.

We define p numeric functions $f_1, \ldots, f_p$ by:

$$f_i(x) = x + 2i/(2p-1), i = 0, \ldots, (p-1).$$

Again we define $F_p(X) = \bigcup_{i=0,\ldots,(p-1)} f_i(X)$ and C_p as the unique solution of $F_p(X) = X, X \subset [0,1]$.

We note that we can build an homeomorphism from $\mathbb{Z}_p$ to C_p by:

$$\sum_{i=0}^{\infty} a_i p^i \to \sum_{i=0}^{\infty} \frac{2a_i}{(2p-1)^{i+1}}.$$

Thus we need now only to find a way to build an homeomorphism between C_p and **C**.

We can, on the other hand, build an homeomorphism between $\mathbb{Z}_p$ and the space of infinite sequences from numbers ranging from 0 to $p-1$ (that is to say: $p^{\mathbb{N}}$):

$$\mathbb{Z}_p \approx \prod_{i=0}^{+\infty}\{0,1,\ldots,p-1\},$$

the homeomorphism being obviously defined by

$$\sum_{n=0}^{n=\infty} a_n p^{-n} \to \{a_n\}_{n\in\mathbb{N}}.$$

The set $\{0,1,\ldots,p-1\}$ being provided with the discrete topology and p^{∞} being provided with the product topology.

1) For binary sequences not infinitely "padded" with 1's, we should write something like this:

$$\frac{1}{2}\sum_{n=0}^{n=\infty} a_n 2^{-n} = \frac{1}{P}\sum_{n=0}^{n=\infty} b_n p^{-n}.$$

This leads to the following calculus.
If $x = \sum_{n=0}^{n=\infty} a_n 2^{-n}$, then:

$$b_N = [p^N \times (\frac{px}{2} - b_0 - b_1 p^{-1}\ldots - b_{N-1} p^{-(N-1)})] \ ^{75},$$

so that b_n will never be chosen as a sequence infinitely padded with $(p-1)$'s if x is of the form $\frac{r}{p^s}$) and $\{b_n\}_n = \varphi(\{a_n\}_n$.

2) For binary sequences of the form $a_1 \ldots a_N 01111\ldots$ that represent a number $\frac{r}{2^s}$ we map them to a (unique) sequence $b_1 \ldots b'_N 0 (p-1)(p-1)(p-1)(p-1)\ldots$ that represents the number $\frac{r'}{p^{s'}}$ defined from $\frac{r'}{p^{s'}} = \varphi(\frac{r}{2^s})$ by the application as shown in Fig. 14.

We order the numbers $\frac{r}{2^s}, r \neq 0(2)$ and $\frac{r'}{p^{s'}}, r' \neq 0 \ (p)$ lexicographically in (s,r) (resp (s',r')). If we note $++$ the iterator operator in these

---

[75] $[x]$ is the integer part of $x \in \mathbb{R}$

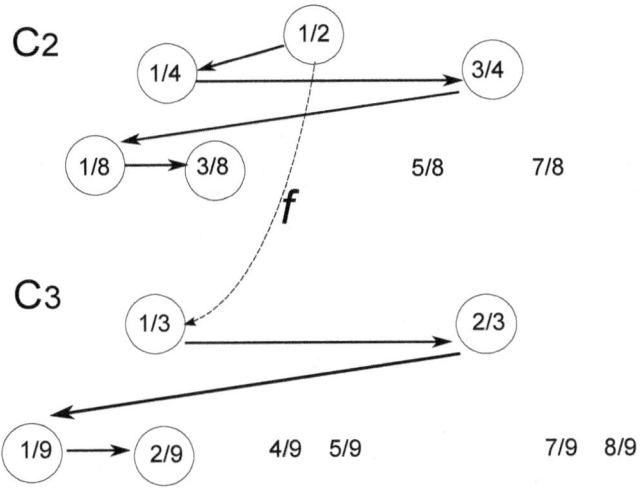

Figure 14: mapping $\frac{r}{2^s}$ to $\frac{r'}{2^{s'}}$, example with p=3

lexicographic orders, we can map each (s,r) to a $f((s,r)) = (s',r')$ so that $f((s,r)++) = f(s,r)++$ (see Fig 14) — that is to say we can go through the lexicographic order in the binary array and go similarly through the lexicographic order in the p-array.

Then we define $\varphi(a_1 \ldots a_N 01111\ldots) = (b_1 \ldots b'_N 0(p-1)(p-1)(p-1)(p-1)\ldots)$ where $(b_1 \ldots b'_N 0(p-1)(p-1)(p-1)(p-1)\ldots)$ is the (only) "infinite" p-adic representation of $\frac{r'}{p^{s'}}$, $(s',r') = f(s,r)$.

So, our application maps the "points of splitting" from C_2 to C_p and maps the others points using their common representation in $[0,1]$. This is therefore a bijective application. The application φ will be continuous in the points that are *not* infinitely padded with 1's and φ^{-1} will be continuous at the points that are not infinitely padded with $(p-1)$'s.

Unfortunately, if we have a closer look, φ will not be continuous at the points infinitely padded with 1's (e.g. that represent $\frac{r}{2^s}$).

The final try is not to build directly an homeomorphism between C and C_p but on the other hand to build an homeomorphism between $[0,1] - C$ and $[0,1] - C_p$ by using the set of complementary intervals.

Solution: We use a construction provided by Svetlana Katok [?] [76] .

We note that the sets C_p are compact (by Tychonoff's theorem , since they are homeomorphic to the product of compact sets), they are also perfect and totally disconnected . So we will use the following result that claims that *all compact, perfect and totally disconnected subsets A of $\mathbb{R}$ are homeomorphic to the Cantor Set C* .

Proof of this result.

We consider such a set A: A is compact in $\mathbb{R}$, therefore it is bounded. A is totally disconnected thus it is nowhere dense (otherwise, we could find a non-trivial interval of $\mathbb{R}$ belonging to A, which is impossible since this would mean that we can find non-trivial connected component in A). We consider $m = \inf(A)$ and $M = \sup(A)$. We build a strictly monotonous application φ from $[m,M]$ into $[0,1]$ such that $\varphi(A) = C$:

---

[76] I am indebted to Eric Chopin for having brought this construction to my attention

A is perfect so $[m,M]-A$ is at most the enumerable union of disjoint intervals (without any common endpoints). A is dense so this union of intervals cannot be but enumerable and not finite.

Let $[m,M]-A = \bigsqcup_{I \in \mathfrak{I}} I$.

We define I_1 as one of the interval of $\mathfrak{I}$ that has the maximal length (by definition, there are only a finite number of intervals having maximal length because $d(I) \leq M-m$, so if d_1 is the maximal length, there are at most $(M-m)/d_1$ intervals with maximal lengths - since they are all disjoint).

We define φ as a monotonous surjective application from I_1 into $[\frac{1}{3}, \frac{2}{3}]$.

We consider $[m,M]-A-I_1 = \bigsqcup_{I \in \mathfrak{I}-I_1} I$. This set is the disjoint union of two enumerable sets of intervals $\bigsqcup_{I \in \mathfrak{I}^{(21)}} I \sqcup \bigsqcup_{I \in \mathfrak{I}^{(22)}} I$.

Again, we can define I_{21} and I_{22} as respective intervals of maximal length for the two subsets (respectively $\mathfrak{I}^{(21)}$ and $\mathfrak{I}^{(22)}$). And we can define φ as a monotonous surjective application from I_{21} into $[\frac{1}{9}, \frac{2}{9}]$ and I_{22} into $[\frac{7}{9}, \frac{8}{9}]$.

Again, we can consider to iterate this process and splitting $\mathfrak{I}^{(21)}$ with I_{21} and $\mathfrak{I}^{(22)}$ with I_{22} that will lead to 4 enumerable set of disjoint intervals $\mathfrak{I}^{(41)}, \mathfrak{I}^{(42)}, \mathfrak{I}^{(43)}, \mathfrak{I}^{(44)}$, etc... Because the length of the intervals in the sets $\mathfrak{I}^{(2^i j)}$ tends toward 0, $[m,M]-A$ is exhausted by the process.

By construction, φ is a monotonous application from $[m,M]-A$ to $[0,1]-C$.

It is rather straightforward to see that φ is a bicontinuous bijection between $[m,M]-A$ and $[0,1]-C$. We extend continuously φ to A by putting:

$$\varphi(a) = \sup\{f(x); x \in [m,M]-A; x < a\}.$$

φ is then a bicontinuous mapping from A into C_p.

In the special case where $A = C_p$, we can define a precise algorithm in order to define φ.

The principle of the algorithm is the following: we will build a set of "B"-objects $\{B_n\}_{n=1}^{n=N}$ with the following properties:

- Each B_n is populated by n_k intervals containing elements of C_p and labeled as "first element, "second element", etc...

- Each B_n is linked to a complementary interval Δ_n to the set C_p.

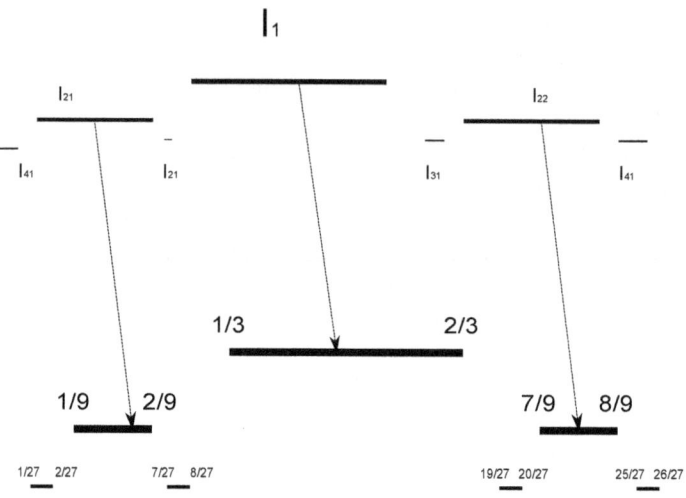

Figure 15: Building of the bijection between $[m, M] - A$ and $[0, 1] - C$

- The objects B_n form a binary tree: each B_n has a father and two children, labeled "0" and "1" and they are also B-objects.

Then we will start with a root object B_1 that have two children: the child "0" is B_2 and the child "1" is B_3. B_2 will contain the interval $[0, 1/(2p - 1)]$ and B_3 will contain the intervals $[2/(2p - 1), 3/(2p - 1)], [4/(2p - 1), 5/(2p - 1)], \ldots, [(2p - 2)/(2p - 1), 1]$.

From here, we will apply the following algorithm:

1) If an object B_n contains at least $d \geq 2$ intervals, then we create 2 new child objects: B_n^0 and B_n^1 (resp. child "0" and child "1").

We populate B_n^0 with the first element of B_n then we populate B_n^1 with the remaining $d - 1$ elements of B_n.

We next consider the interval Δ_n defined as the interval separating the first and the second element of B_n and we associate Δ_n to B_n.

We set the object B_n to "non-active" state so that we flag it to have been already processed.

2) If an object B_n only contains one element I, we split this interval element into p new interval elements $I_1, I_2, \ldots I_p$ obtained by the splitting process used to build the cantor set C_p. That is to say the interval I is divided into $2p - 1$ intervals of equal length and $I_1, \ldots I_p$ are the intervals that are complementary to C_p.

From here, we create again two children B_n^0 and B_n^1 but we populate the first one with I_1 and the other one with remaining $p - 1$ intervals. e.g. $I_2, \ldots I_p$. The process is after this, similar to 2).

After an iteration of the algorithm, we will have created new objects B from the existing one, then after n iterations of the algorithm, we will build a tree of B Objects with N components.

In order to build the bijection between C_2 and C_p we will associate the binary tree of complementary intervals of C_2 and the binary tree of B-objects.

Then we consider f as the function from objects B to intervals objects defined by $f(B) = \Delta$. If we label the elements of the binary tree by a binary sequence $s \in 2^N$ the bijection φ is defined by:

$$\varphi(I_s) = f(B_s).$$

In fact if we note $B_{p,s}$ the sth B set for C_p, we have $I_s = f(B_{2,s})$ [77].

The final step is to browse the binary tree of the complementary intervals of C_2, using for example base 2 numbering of the elements so that is $s = 1a_1 a_2 \ldots a_n$ is binary sequence, the element B_s can be found by using the following "genealogy": we start with B_1, then we take the child a_1, then the child a_2 of a_1, etc... (See Fig. 16).

If we define the continuous functions φ_N from $[0, 1]$ into $[0, 1]$ by:
$\varphi_N(f(B_{2,i})) = f(B_{p,i})$ for $1 \leq i \leq 2^N$ (that is to say that we descend into the B-tree with a level of N)

then $\lim_{N \to \infty} \varphi_N = \varphi$. We can implement the algorithm with the Perl language (see the appendix for the source code) and we get a graphical representation of φ_N (See Fig. 17).

---

[77] So that we also may read $\varphi \circ f(B_{2,s}) = f(B_s)$

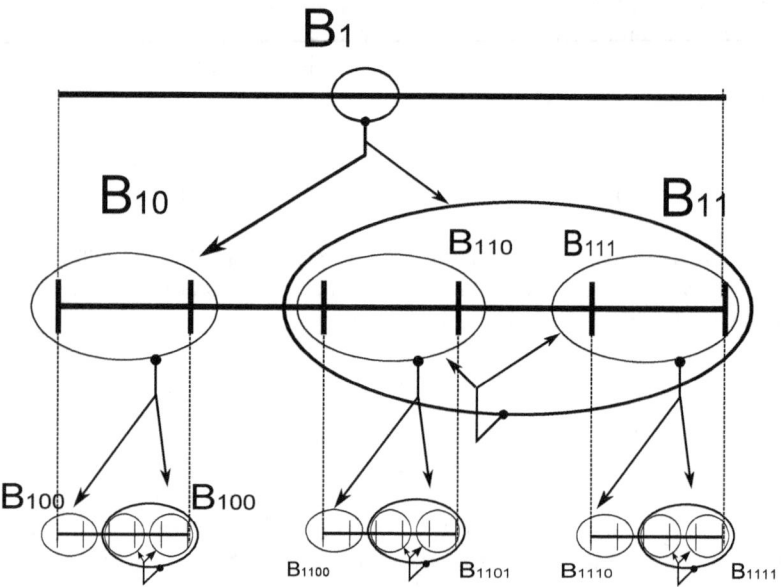

Figure 16: Algorithm creating the bijection φ between C_2 and C_p when p=3

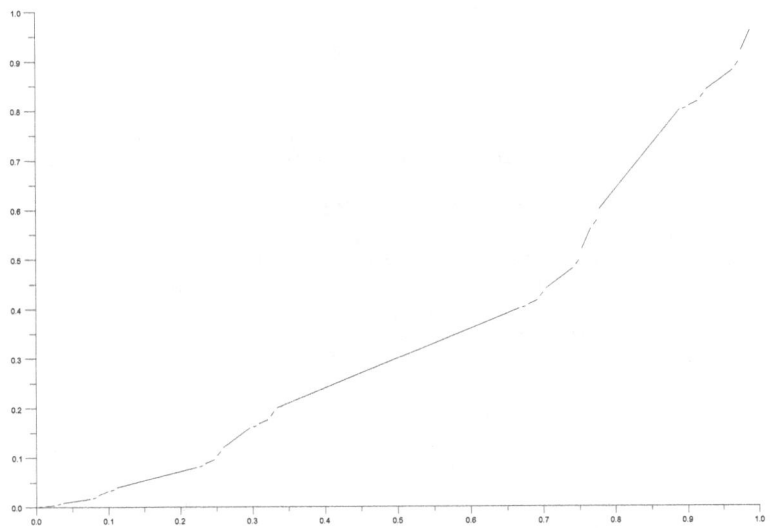

Figure 17: Graphical representation of the homeomorphism φ_5 between C_2 and C_3 with a level of recursion =5

Notes

[i]A process due to Alexandrov [?] allows us to build a continuous surjection φ from C into every nonempty compact metric space E. We use a family of partitions of C by clopens sets[78], say: $\{\{X_{n,k}\}1 \leq k \leq m(n)\}$ and a family of covering of E by closed non empty sets, say: $\{\{Y_{n,k}\}1 \leq k \leq m(n)\}$ such that the following properties hold:

1) The diameter of $X_{n,k}$ is $< \frac{1}{n+1}$, for any k.

2) The diameter of $Y_{n,k}$ is $< \frac{1}{n+1}$, for any k.

3) The family $\{\{X_{n+1,k}\}1 \leq k \leq m(n+1)\}$ is included in the family $\{\{X_{n,k}\}1 \leq k \leq m(n)\}$.

4) For any k,l such that $1 \leq k \leq m(n)$ and $1 \leq l \leq m(n+1)$ then:

$$X_{n+1,l} \subset X_{n,k} \Rightarrow Y_{n+1,l} \subset Y_{n,k}.$$

Then we can see that every sequence $\{k(n)\}_{n \in \mathbb{N}}$ defines uniquely an element $x \in C$ by $x = \bigcap_{n \in \mathbb{N}} X_{n,k(n)}$, and we can check that the corresponding sequence of $Y_{n,k(n)}$ defines a unique point $y \in E$ by $y = \bigcap_{n \in \mathbb{N}} Y_{n,k(n)}$. We define a mapping φ from C into E by making $\varphi(x) = y$ and we can check that this mapping is continuous and surjective.

In the case where $C = C_2$ and $E = C_3$ (or C_p), we can choose the families $X_{n,k}$ and $Y_{n,k}$ by an identical process than the process we have described before: The family $X_{n,k}$ will be the family of *remaining* intervals after n steps, $1 \leq k \leq 2^n$, and the family $Y_{n,k}$ will be picked from the family or remaining intervals of C_3 as shown in the following figure (Fig. 18):

The function φ will be a bijection and therefore will define an homeomorphism. More precisely, φ will be defined for every sequence $\{k(n)\}_{1 \leq k(n) \leq 2^n, n \in \mathbb{N}}$ by:

$$\bigcap_{n \in \mathbb{N}} X_{n,k(n)} \rightarrow \bigcap_{n \in \mathbb{N}} Y_{n,k(n)}.$$

If we define the function φ_N by:

$$\bigcup_{1 \leq k \leq 2^N} X_{N,k} \rightarrow [0,1],$$

$$X_{N,k} \rightarrow \varphi_N(X_{N,k}) = Y_{N,k}.$$

Then the function $\varphi = \lim_{N \to \infty} \varphi_N$ will be the required homeomorphism.

The idea is moreorless to embedd a "p-covering" into a "2-covering", and this is equivalent to the process detailed in the solution of this Exercise.

[ii]We have the following homeomorphisms $\mathbb{Z}_p \approx p^\mathbb{N} \approx C_p \approx C_2 \approx 2^\mathbb{N} \approx \mathbb{Z}_2$.

[iii]There are infinitely many ways to build the homemomorphisms φ_N in the way we have described. It is not difficult to see that the set of homeomorphisms has the power of the continuum.

[iv]A perl source code which build the applications φ_n is available in the appendix section.

---

[78]E.g. sets that are both closed and open for the topology of C

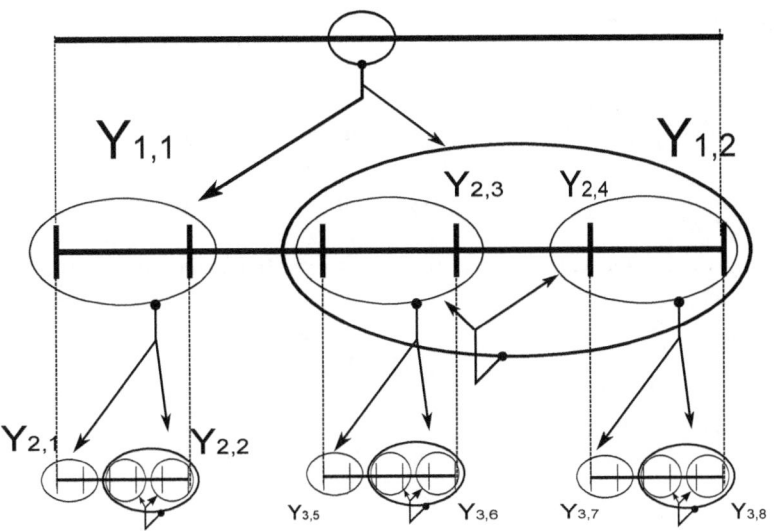

Figure 18: Family $Y_{n,k}$ if $E = C_3$

43/828 ¶

Subject: p-adic sign

Show that, for every $x \in \mathbb{Z}_p$, we can define the p-adic sign by:

$$sgn_p(x) = \lim_{n \to \infty} x^{p^n}.$$

Show that the values of the p-adic sign are 0 and the $(p-1)$ roots of unity in $\mathbb{Q}_p$.

SOLUTION:

We now show that:

$$\forall n > 0, x^{p^n} = x^{p^{n-1}} + p^n u_n, u_n \in \mathbb{Z}_p.$$

For $n = 1$, we have, following Fermat's little theorem:

$$x^{p-1} = 1 + kp, k \in \mathbb{Z}.$$

This implies:

$$x^p = x + p \times u_1, u_1 = kx \in \mathbb{Z}.$$

Let us suppose that the property we want to prove is true for $1, \dots, n$, then:

$$
\begin{aligned}
x^{p^{n+1}} &= (x^{p^{n-1}} + p^n u_n)^p, u_n \in \mathbb{Z}_p, \\
&= x^{p^n} + pp^n u_n x^{(p-1)p^{n-1}} + (p(p-1)/2)p^{2n} u_n^2 x^{(p-2)p^{n-1}} + \ldots + p^{pn} u_n^p, \\
&= x^{p^n} + p^{n+1} u_{n+1}, u_{n+1} \in \mathbb{Z}_p.
\end{aligned}
$$

So that this property is true for all $n > 0$.
We can then write $x^{p^n} = x + \sum_{i=1}^{i=N} p^i u_i$ and, hence:

$$
sgn_p(x) = x + \lim_{n \to \infty} \sum_{i=1}^{i=n} p^i u_i.
$$

The leading term of the sum is such that $||p^i u_i||_p \le p^{-i}$ so that the sum is convergent[79] and then $sgn_p(x)$ can be defined.
On the other hand, we have:

$$
sgn_p(x) - x = \sum_{i=1}^{+\infty} p^i u_i.
$$

So that:

$$
||sgn_p(x) - x||_p \le \frac{1}{p}.
$$

We have also, if $sgn_p(x) = sgn_p(y), x \ne y$:

$$
\begin{aligned}
||x - y|| &= ||sgn_p(x) - x - (sgn_p(y) - y)||_p, \\
&\le ||sgn_p(x) - x||_p + ||sgn_p(y) - y||_p, \\
&\le \frac{2}{p}.
\end{aligned}
$$

If $(x,y) \in \{0, \ldots, p-1\}$, then $||x - y||_p = 1$ so that, when x describes $\{0, \ldots, p-1\}$, the values of $sgn_p(x)$ are all distinct.

---

[79] A criterion for $\sum_{n \ge 0} x_n$ to be convergent in an ultrametric space is that $||x_n|| \to 0$ when $n \to \infty$

From the definition of sgn_p, we note that:

$$sgn_p(x)^p = (\lim_{n\to\infty} x^{p^n})^p = \lim_{n\to\infty} x^{p^{n+1}} = sgn_p(x).$$

So either we have $sgn_p(x) = 0$ or sgn_p is a root of $X^{p-1} - 1 = 0, X \in \mathbb{Q}_p$.
The polynomial $P(X) = X^{p-1} - 1$ has at most $p - 1$ roots in $\mathbb{Q}_p$ so that
the $p - 1$ power roots of unity, Φ_{p-1}, and 0 describe the p-adic signs:

$$\{sgn_p(x), x \in \mathbb{Z}_p\} = \Phi_{p-1} \cup \{0\}.$$

44/828 ¶

Subject: Domain of definition of two p-adic functions

Find, in $\mathbb{Q}_p$ the domain of definition of the following functions:

$$x \to \sum_{k=0}^{+\infty} \frac{x^k}{k!} \quad \text{and} \quad x \to \sum_{k=0}^{+\infty} (-1)^{k-1} \frac{x^k}{k}.$$

SOLUTION:

If we have, for $k \in \mathbb{N}$, $k = p^j \times m$, $m \geq 1$ then $||k||_p = p^{-j}$.

We have that $k \geq p^j$ or equivalently $j \leq \frac{\ln(k)}{p}$ and we may write:

$$||k||_p \geq p^{-\frac{\ln(k)}{p}}.$$

We use this to get:

$$
\begin{aligned}
||\frac{x^k}{k}||_p &= \frac{(||x||_p)^k}{||k||_p}, \\
&\leq (||x||_p)^k . p^{\frac{\ln(k)}{p}} \quad \text{(see properties of the p-adic norm in Exercise 37),} \\
&\leq (||x||_p)^k . k^{\frac{\ln(p)}{p}}.
\end{aligned}
$$

The functions of type $k \to exp(\alpha k), \alpha > 0$ dominate[80] the functions of type $k \to k^\beta . \beta > 0$ when $k \to \infty$, so that for $||x||_p < 1$ $(||x||_p)^k (k^{\ln(p)})^{\frac{1}{p}}$ tends towards 0 when $k \to \infty$ [81].

On the other hand, if $||x||_p \geq 1$, $||\frac{x^k}{k}||_p$ will tend toward infinity, for $k \to \infty$, since $||k||_p \leq 1 \Rightarrow ||\frac{x^k}{k}||_p \geq ||x||_p^k$.

This means that the sum $\sum_{k=0}^{+\infty}(-1)^{k-1}\frac{x^k}{k}$ will converge for and only for the $x \in \mathbb{Q}_p$ such that $||(-1)^{k-1}\frac{x^k}{k}||_p \to 0$ when $k \to \infty$, that is to say, for the x such that $||x||_p \geq 1$ [82].

For the other function, we can write:

$$k! = k(k-1)(k-2)\ldots 2 \times 1.$$

In the set $\{1, 2, \ldots, k-2, k-1, k\}$ there is exactly $[\frac{k}{p}]$ multiples of p so that:

$$p^{-[\frac{k}{p}]} \mid k!.$$

Again, in the set $\{1, 2, \ldots, k-2, k-1, k\}$ there is exactly $[\frac{k}{p^2}]$ multiples of p^2 so that:

$$p^{-[\frac{k}{p^2}]} \mid \frac{k!}{p^{[\frac{k}{p}]}}.$$

And, recursively, for every $m > 0$, in the set $\{1, 2, \ldots, k-2, k-1, k\}$ there is exactly $[\frac{k}{p^m}]$ multiples of p^m so that:

$$p^{-[\frac{k}{p^m}]} \mid \frac{k!}{p^{\sum_{0<i<m}[\frac{k}{p^i}]}}.$$

After a certain time, we will exhaust all the multiples of powers of p in the set $\{1, 2, \ldots, k-2, k-1, k\}$ and we can write:

---

[80] See Exercise 3

[81] $||x||_p^k k^{\frac{\ln(p)}{p}} = \frac{k^\beta}{\exp(\alpha k)}, \beta = -\frac{\ln(p)}{p}, \alpha = \ln(\frac{1}{||x||_p}) > 0$

[82] A criterion for a sum $\sum_n a_n$ to converge in a ultrametric space is simply that $a_n \to 0$ when $n \to \infty$

$$k! = a.p^{\sum_{i=1}^{+\infty}[\frac{k}{p^i}]}, \Delta(a,p) = 1 \ ^{83}.$$

Using this last result, we get:

$$||k!||_p \geq p^{-\sum_{i=1}^{+\infty}[\frac{k}{p^i}]} \geq p^{-\frac{k}{p-1}}$$

so that:

$$||\frac{x^k}{k!}||_p \leq (||x||_p)^k.p^{\frac{k}{p-1}} \leq (||x||_p p^{\frac{1}{p-1}})^k.$$

If $||x||_p < (\frac{1}{p})^{\frac{1}{p-1}}$, the term $||\frac{x^k}{k!}||_p$ will tend toward 0 when k tends toward infinity.

On the other hand, if $||x||_p \geq (\frac{1}{p})^{\frac{1}{p-1}}$, then we have again:

$$||\frac{x^k}{k!}||_p \geq p^{\sum_{i=1}^{+\infty}[\frac{k}{p^i}]-\frac{k}{p-1}}$$

for the subsequence $k_N = p^N$, $||\frac{x^{k_N}}{k_N!}||_p \geq p^{\frac{-1}{p-1}} > 0$, for $N \to \infty$ so that the function will not converge.

For $p = 2$, the domain $\{x \in \mathbb{Q}_2, ||x||_2 < \frac{1}{2}\}$ is $4\mathbb{Z}_2$ and for $p > 2$, the domain $\{x \in \mathbb{Q}_p, ||x||_p < (\frac{1}{p})^{\frac{1}{p-1}}\}$ is the $\mathbb{Q}_p$-unit ball, e.g. $\{x, ||x||_p < 1\}$. Indeed, the set $\{||x||_p, x \in \mathbb{Q}_p\}$ is discrete $= \{p^{-j}, j \in \mathbb{Z}\}$. If $||x||_p < 1$, then $||x||_p = p^{-j}, j \geq 1 \Rightarrow ||x||_p < p^{-\frac{1}{p-1}}$ since $j > \frac{1}{p-a}, \forall j, \forall p > 2$. If $||x||_p < p^{-\frac{1}{p-1}}$ this implies $||x||_p = p^{-j}, j > \frac{1}{p-1} \Rightarrow j \geq 1 \Rightarrow ||x||_p < 1$.

---

[83] In fact we could also write, more precisely: $k! = a.p^{\sum_{i=1}^{\frac{\ln(||k||_p)}{\ln(p)}}[\frac{k}{p^i}]}...$

45/828 ¶

Subject: Orders in $\mathbb{Q}_p$

Show that, in $\mathbb{Q}_p$ there is no order relationship $\prec$ that satisfies to the following conditions:

a) $x \succ 0$ and $y \succ 0 \implies x+y \succ 0$;

b) $x \succ 0$ and $y \succ 0 \implies xy \succ 0$;

c) $x_n \succ 0$ and $\lim_{n\to\infty}(x_n) = x \implies x \succeq 0$.

SOLUTION:

Before we use the hint, we need to check that if $\succ$ is such an order in $\mathbb{Q}_p$, then it will coincide necessarily over $\mathbb{Q}$ with the usual order: since $\forall (x,y) \in \mathbb{Q}^2, x \succ 0, u \succ 0 \Rightarrow x+y \succ 0$, we have $x \prec 0 \Rightarrow -x \succ 0$ and $x \succ 0 \Rightarrow -x \prec 0$. If $1 \prec 0$, then $-1 \succ 0$ and $(-1)^2 \succ 0$. So that we must have $1 \succ 0$. If $x \in \mathbb{N}$ such that $x \succ 0$, then $x+1 \succ 0$ from the properties of $\succ$ and this proves that the order $\succ$ coincides with the usual order $>$ over $\mathbb{N}$. If $-x \in -\mathbb{N}$ then $x \succ 0$ and this implies $-x \prec 0$. This shows that $\succ$ coincides with the usual order over $\mathbb{Z}$. Let us now consider $p \succ 0$ and $q \succ 0$. If $\frac{p}{q} \prec 0$ then $q \times \frac{p}{q} \prec 0$ that is to say $p \prec 0$, which is impossible, so we must have $p \succ 0$ and $q \succ 0 \Rightarrow \frac{p}{q} \succ 0$. We use the same method for the other cases $(p \prec 0, q \succ 0), (p \succ 0, q \prec 0)$ and $(p \prec 0, q \prec 0)$. Finally we get that $\succ$ coincides with the usual order over $\mathbb{Q}$.

The positive sequence

$$x_n = \sum_{i=0}^{i=N} (p-1)p^n$$

converges toward -1 in $\mathbb{Q}_p$: $\sum_{i=0}^{+\infty}(p-1)p^n = (p-1)\frac{1}{1-p} = -1$
so that condition c) is not satisfied.

Notes

[i] $\mathbb{Q}_p$ is isomorph to the product of $\mathbb{Z}$ copies of $\{0,1,\dots,p-1\}$, e.g. $\mathbb{Q}_p \approx \prod_{\alpha\in\mathbb{Z}}\{0,1,\dots,p-1\}$ but $\mathbb{Z}$ is not well ordered (by its natural order) so we cannot use the exercise 13 to provide a total (lexicographical) order in $\mathbb{Q}_p$, but we can do it for $\mathbb{Z}_p$, indeed: $\mathbb{Z}_p \approx \prod_{\alpha\in\mathbb{N}}\{0,1,\dots,p-1\}$.

46/828

Subject: Propertics of the representation of numbers in base 10 and p-adics

We define a distance δ in $\mathbb{N}$ by $\delta(m,n) = \frac{1}{k}$ if m and n both terminates with the same k last decimals[84]. Example: $\delta(1233332122, 51332122) = \frac{1}{6}$.

a) Show that $(\mathbb{N}, \delta)$ is not complete and that its completion is isomorph (as a ring) to the direct product $\mathbb{Z}_2 \times \mathbb{Z}_5$.

b) Prove that, for every number k, there is exactly 4 k–digits motives that replicate themselves by multiplication:

if:

$$N_1 = A \ldots \boxed{a_1 a_2 \ldots a_k} \ldots$$
$$N_2 = B \ldots \boxed{a_1 a_2 \ldots a_k} \ldots$$

then:

$$N_1 N_2 = C \ldots \boxed{a_1 a_2 \ldots a_k} \ldots.$$

SOLUTION:

First we note that if $\{m_n\}$ is a Cauchy sequence in $(\mathbb{N}, \delta)$, then

$$m_p - m_q = 10^{\varphi(p,q)}(A - B)$$

and $\varphi(p,q) \to \infty$ when $(p,q) \to \infty$ (because $\delta(m_p, m_q) = \frac{1}{\varphi(p,q)}$)

---

[84]The $k-1$th decimals being different

that means that:

$$||m_p - m_q||_2 \leq 2^{-\varphi(p,q)}$$

and:

$$||m_p - m_q||_5 \leq 5^{-\varphi(p,q)}.$$

So that $\{m_n\}$ is also a Cauchy sequence in both $\mathbb{Z}_2$ and $\mathbb{Z}_5$.
Conversely, if $\{m_n\}$ is a Cauchy sequence in $\mathbb{Z}_2 \times \mathbb{Z}_5$:

$$||m_p - m_q||_2 = 2^{-\varphi(p,q)}$$

and:

$$||m_p - m_q||_5 = 5^{-\psi(p,q)}$$

and $\varphi(p,q) \to \infty, \psi(p,q) \to \infty$ when $(p,q) \to \infty$.
We can write:

$$m_p - m_q = 10^{-\min(\varphi,\psi)(p,q)}C$$

and consequently:

$$d(m_p, m_q) \leq \frac{1}{\min(\varphi, \psi)(p,q)}.$$

This shows that there is a total equivalence between Cauchy sequences in $(\mathbb{N}, \delta)$ and in $\mathbb{Z}_2 \times \mathbb{Z}_5$. Let $\widetilde{\mathbb{N}}$ be the completion of $(\mathbb{N}, \delta)$.
We build the following application Φ from $\widetilde{\mathbb{N}}$ into $\mathbb{Z}_2 \times \mathbb{Z}_5$:
if $\{x_n\}$ is a Cauchy sequence in $\mathbb{Z}_2 \times \mathbb{Z}_5$, then if

$$x_n \to x \in \widetilde{\mathbb{N}},$$

we put:

$$\Phi(x) = \lim_{x_n \in \mathbb{Z}_2 \times \mathbb{Z}_5} x_n$$

Φ is an isomorphism so that we have:

$$\widetilde{\mathbb{N}} \approx \mathbb{Z}_2 \times \mathbb{Z}_5.$$

b) if we consider the sequence $A_k = \boxed{a_1 a_2 \ldots a_k}$ where $\boxed{a_1 a_2 \ldots a_k}$ is the k-motive we need to find, the the $\{A_k\}_k$ converges in $\widetilde{\mathbb{N}}$, let A be its limit, then we have:

$$A^2 = A, A \in \widetilde{\mathbb{N}}.$$

If we transport this equation in $\mathbb{Z}_2 \times \mathbb{Z}_5$, we see that A is represented by the product of the solutions of $x^2 = x$ in $\mathbb{Z}_2$ with the solutions of $x^2 = x$ in $\mathbb{Z}_5$ that is to say that:

$$\{A\} \approx \{(0,0), (0,1), (1,0), (1,1)\}.$$

1) if A is represented by $(0,0)$ in the ring $\mathbb{Z}_2 \times \mathbb{Z}_5$, then:

$$A \equiv 0 \ (2^N), A \equiv 0 \ (5^N), \forall N > 0.$$

This is only verified by $A = \ldots 0 \ldots 0000$.

2) if A is represented by $(0,1)$ in the ring $\mathbb{Z}_2 \times \mathbb{Z}_5$, then:

$$A \equiv 0 \ (2^N), A \equiv 1 \ (5), A \equiv 0 \ (5^{N+1}), \forall N > 0.$$

We must have:

$$A = 10R + a_0 \equiv 0 \ (2), A = 10R + a_0 \equiv 1 \ (5), 0 \le a_0 \le 9,$$

this leads to $a_0 = 6$.

$$A = 100R + 10a_1 + 6 \equiv 0 \ (4), A = 100R + 10a_1 + 6 \equiv 1 \ (25), 0 \le a_1 \le 9,$$

this leads to $a_0 = 7$.

We have an algorithm that allow us to compute $A = \ldots 109376$.

3) if A is represented by $(1,0)$ we proceed the same way and we find that $A = \ldots 890625$.

4) if A is represented by $(1,1)$ we proceed the same way and we find that $A = \ldots 000001$.

Exercises 47 to 69
Categories and Functors

"There was also some fun with the choice of the terminology. Since the philosopher Kant had made ample use of general categories, the term was borrowed from him for its present mathematical use [...] Carnap in his book "on Die Logische Syntax Der Sprachen" had talked of functors in a different sense and made some corresponding mistakes. It seemed in order to take that word for a better and less philosophical purpose".

(In "a Geometrical Point of View: A Study of the History and Philosophy of Category Theory", Jean-Pierre Marquis)

47/828 $\mathscr{B}$

Subject: Universal Repelling object in $\mathscr{P}(X)$

Let $\mathscr{P}(X)$ be the power set of a fixed set X. We define a category $\mathscr{C} = (\mathscr{P}(X), \{i\})$ where the morphisms between two parts A and B are being described by the canonical injection[85] $i_{A \to B} : x \to x$ if $A \subset B$ and by $\emptyset$ otherwise[86]. Show that this category does not have a universal repelling object but that the dual category has such an object[87].

SOLUTION:

If $\mathscr{C}$ would have a universal repelling object Z, then we could find Z such that $Z \subseteq X$ and such that Z is linked to every part of $A \subseteq X$ by a unique injection:

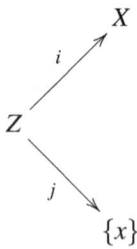

---

[85] We refer to an injection as an injective mapping

[86] E.g. if $A \not\subseteq B$, then $Mor(A, B) = \emptyset$

[87] Since the definition of the morphisms in this set was unclear to us, we have chosen this interpretation

We know that the *singleton* $\{0\}$ is always an element of $\mathscr{P}(X)$ but we may not find an injection from Z into 0. We conclude that there is no universal objject.

In order to find a universal object in the dual category $\mathscr{C}^0$, we must find a part Z such that Z is reached by all the other parts A of X by a unique injection:

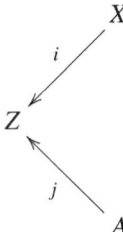

If we take $Z = X$ then, for every part $A \subset X$, there exists trivially only one canonical injection $i_A \colon A \to X$.

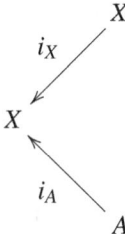

This proves that X is the universal object of the dual category $\mathscr{C}^0$.

48/828 $\mathscr{B}$

Subject: Contravariant functor in the categories of the parts of a given set

Build a contravariant functor from the Category $\mathscr{C}$ of the parts of a given set (see Ex 46) into itself.

SOLUTION:

Let us define the functor F , $\mathscr{C} \to \mathscr{C}$ by:

$$F(A) = X \backslash A,$$
$$F(i_{A,B}) = i_{X \backslash B, X \backslash A} , i_{A,B} \in Mor(A,B).$$

Where $i_{A,B}$ is the canonical injection defined by:

$$A \to B, A \subseteq B,$$
$$x \to x.$$

F is a contravariant functor of $\mathscr{C}$ into itself.
If we have the following commutative diagram (where $A \subset B \subset C$):

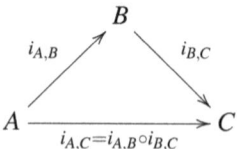

Then F will transform it into the following commutative diagram:

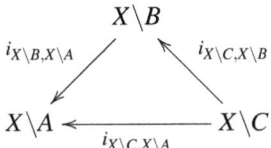

We have:

$$
\begin{aligned}
F(i_{B,C} \circ i_{A,B}) &= F(i_{A,C}), \\
&= i_{X\setminus C, X\setminus A}, \\
&= i_{X\setminus B, X\setminus A} \circ i_{X\setminus C, X\setminus B}, \\
&= F(i_{A,B}) \circ F(i_{B,C}).
\end{aligned}
$$

What shows that F is a contravariant functor.

49/828 $\mathscr{B}$

Subject: Universal repelling object in the category of groups and in the category of vector spaces

Can we find a universal repelling object in the category of the groups? In the category of vector spaces of a given field $\mathbb{K}$? In the dual categories of the two mentioned spaces?

SOLUTION:

If we consider the category $\mathscr{C}_G$ of groups where the morphisms are the homomorphisms between two groups, then the trivial group $\{e, \star\}$ — where e is a unit for $\star$ — is a universal repelling object:

$$
\begin{array}{ccc}
 & (B, +) & \\
 \varphi_{e,B} \nearrow & & \\
 \{e, \star\} \xrightarrow[\varphi_{e,A}]{} & (A, *) &
\end{array}
$$

Indeed, for every group $A = (A, *)$, there exists only one homomorphism $\varphi_{e,A}$ from $\{e, \star\}$ onto A defined trivially by $\varphi_{e,A}(e) = e_A$. e_A being the unit element of $(A, *)$.

If we consider the category $\mathscr{C}_V^K$ of vector spaces on a given field $\mathbb{K}$ where the morphisms are the linear mappings, we also can find a universal repelling object by considering the null vector space $\{\mathbf{0}\}$:

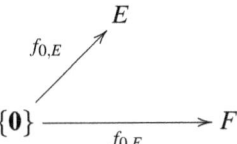

For every vector space E, we can find a unique linear mapping $f_{0,E}$, $\{\mathbf{0}\} \rightarrow E$, defined trivially by $f_{0,E}(0) = 0_E$.

Now, if we consider the dual categories, we also can find in both cases a universal repelling object:

1) Category $\mathscr{C}_G$ of groups.

The monoid $\{e, \star\}$ is also a universal repelling object:

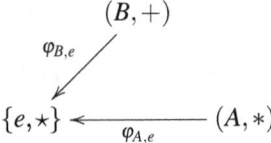

For every group $A = (A, *)$ there exists only one homomorphism $\varphi_{A,e}$: $(A, *) \rightarrow \{e, \star\}$. This homomorphism is defined by $\varphi_{A,e}(x) = e$, $\forall x \in A$.

2) Category $\mathscr{C}_V^K$ of vector spaces on a given field $\mathbb{K}$.

The null vector space $\{\mathbf{0}\}$ is also a universal repelling object:

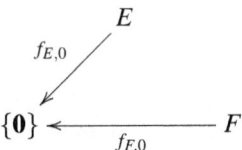

For every vector space E there exists only one linear mapping $f_{E,0}$: $E \rightarrow \{\mathbf{0}\}$. This linear mapping is defined by $f_{E,0}(x) = 0$, $\forall x \in E$.

50/828 $\mathcal{B}$

Subject: Category of Abelian group with distinguished generator

Let G_1 the category of Abelian groups with a distinguished generator (the morphisms being the group homomorphisms mapping the distinguished generator into another distinguished generator). Name the universal objects of G_1 and G_1^0.

SOLUTION:

We have two possible interpretation of the problem statement: 1) we consider the category of Abelian groups with one generator (cyclic groups) or 2) we consider the category of Abelian groups with many generators and we pick up (distinguish) one.

Interpretation 1).

Let our category G_1 be the pairs (f, G) where G is a cyclic group f is an application from the singleton $\{1\}$ into G such that $f(1)$ generates G.

A morphism φ between two objects (f, G) and (h, G') in that category will be a group homomorphism $G \rightarrow G'$ such that the following diagram is commutative:

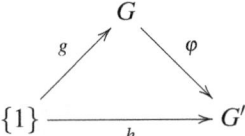

As always we can easily check that this defines a category.
If we consider this diagram:

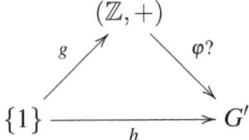

$(\mathbb{Z},+)$ will be the universal object if and only if, given an abelian group
G', we can always find a morphism φ making this diagram commutative.
If we define φ by:

$$\varphi(n) = h(1)^n, n \in \mathbb{Z}.$$

Then we can check that φ is a group-homomorphism and have the required properties. Unicity of φ is guaranteed by the fact that all such group-
homomorphisms with these properties coincide since they must coincide
over the generator. We have proved that $(\mathbb{Z},+)$ is the universal object of
our category.

In the dual category, we need to show that we can always find a unique
group-homorphim φ making the following diagram commutative:

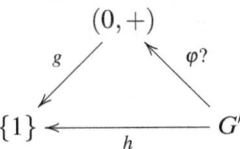

And we may define φ by $\varphi(x) = 0$ for any $x \in G'$. Unicity is obvious and
we immediatly check that φ is a group homomorphism with the required

properties.

Interpretation 2).

Let $K(\alpha)$ and $L(\beta)$ be two objects in G_1. $\alpha \neq \beta$, $\alpha \in S$ and $<S>=K$, $\beta \in T$ and $<T>=L$. A morphism $K(\alpha) \to L(\beta)$ is a group homomorphism $\varphi\colon K(\alpha) \to L(\beta)$ such that: $\varphi(\alpha) = \beta$. We note that in G_1, the group K is not necessarily different to L. Because K and L are Abelian:

$$\forall x \in K, x = \alpha^n \times \prod_{i=1...N, x_i \in S-\alpha} x_i,$$

$$\forall y \in L, y = \beta^{n'} \times \prod_{i=1...N', y_i \in T-\beta} y_i.$$

That is to say: $K(\alpha) =< \alpha > \times < S-\alpha >$, $K(\beta) =< \beta > \times < T-\beta >$. An application $\varphi \in \mathrm{HOM}(K,L)$ given by: $\varphi(\alpha) = \beta$ is such that:

$$\begin{aligned}
\varphi(x) &= \beta^n \times \prod_{i=1...N', \varphi(x_i) \in <T-\beta>} \varphi(x_i) \times \prod_{j=1...N-N', \varphi(x_i) \in <\beta>} \beta, \\
&= \beta^{n'} \times \prod_{i=1...N', y_i \in T-\beta} y_i.
\end{aligned}$$

So that $\varphi \in \mathrm{Mor}(K(\alpha), L(\beta))$.

$\mathbb{Z}(+1)$ is the universal object for G_1:
First we note that $\mathbb{Z}(+1)$ and $\mathbb{Z}(-1)$ are isomorphs because there exist a morphism φ in $\mathrm{Mor}(\mathbb{Z}(+1), \mathbb{Z}(-1))$ defined by $\varphi(x) = -x$ such that $\varphi^2 = 1_{Z(+1)}$.
Trivially, for every $K(\alpha) \in G_1$ there exist only one morphism:

$$\varphi_{K,\alpha} \in Mor(\mathbb{Z}(+1), K(\alpha))$$

defined by:

$$\varphi_{K,\alpha}(n) = \alpha^n$$

and

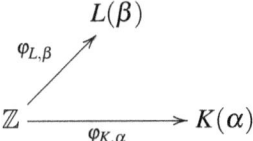

Conversely, the trivial group $(0,+)$ is a universal (repelling) object for the dual category G_1^0: for every $K(\alpha) \in G_1$ there exist only one morphism $\varsigma_{K,\alpha} \in \mathrm{Mor}(K(\alpha),(0,+))$ defined by:

$$\varsigma_{K,\alpha}(n) = 0$$

and

$$
\begin{array}{ccc}
 & L(\beta) & \\
\varsigma_{L,\beta} \swarrow & & \\
(0,+) \xleftarrow{\quad \varsigma_{K,\alpha} \quad} & & K(\alpha)
\end{array}
$$

51/828

Subject: The Free Group with two generators

Let $\mathfrak{C}_2$ the category of groups with two distinguished generators (the morphisms being the group homomorphisms mapping the two distinguished generators into two others distinguished generators). Show that $\mathfrak{C}_2$ have a universal object. This object is called the Free Group with two generators.

SOLUTION:

Prerequisites and discussion.

We recall now the classical construction of free groups (this can be found for example in [?]).

If G is a group and S a set. We consider the category $\mathfrak{C}_S$ made with couples (G, φ) where φ is an application: $S \to G$. The morphisms of this category are defined as follows:

$f \in Mor((G, \varphi), (G', \psi))$ if and only if:

$-f \in Hom(G, G')$,

$-f \circ \varphi = \psi$.

$\mathfrak{C}_S$ contains elements (G, φ) such that φ generates G, e.g. such that $< \varphi(S) >= G$ [88]. In that latter case we note that there can exist at most one element in $Mor((G, \varphi), (G', \psi))$ for any other element (G', ψ) of the

---

[88] We note $< X >$ the group generated by a set X

category. Indeed, this homomorphism is defined as the application $f \in Hom(G, G')$ that makes the following diagram commutative:

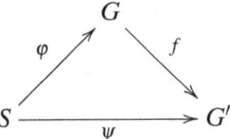

If this homomorphism exists, it is uniquely defined by the parameters $(s_i, a_i)_{i=1}^{i=N}$ (where the s_i's are elements of S and $a_i \in \mathbb{Z} - \{0\}$, $i = 1, \ldots, N$).

$$x = \prod_{i=1,\ldots,N} \varphi(s_i)^{a_i},$$

$$f(x) = \prod_{i=1,\ldots,N} f(\varphi(s_i))^{a_i} = \prod_{i=1,\ldots,N} \psi(s_i)^{a_i}.$$

Which means that if we can find a group G generated by φ such that $Mor(\varphi, \psi) \neq \emptyset$ for all $\psi \in \mathfrak{C}_S$ then (G, φ) *will be* a universal object for the category. This group is then *defined* as the Free Group generated by S. When $S = \{0, 1\}$ this group is the universal object of $\mathfrak{C}_2$ we are looking for.

We could think that, apparently, we can always find such an homomorphism f for any element (G', ψ) and that it is enough to define it by the formula:

$$x = \prod_{i=1}^{i=N} \varphi(s_i)^{\alpha_i}, \ f(x) = \prod_{i=1}^{i=N} \psi(s_i)^{\alpha_i}.$$

For example, if we consider an Abelian group G with 2 generators a and b and φ: $\{0, 1\} \to \{a, b\}$. If G' is a group with 2 generators u and v such that $uv \neq vu$ and ϕ: $\{0, 1\} \to \{u, v\}$ then there cannot exist any morphism in $f \in Mor(\varphi, \psi)$ for this would mean that $f(ab) = f(ba)$ or $f(a)f(b) = f(b)f(a)$ or $uv = vu$, which is impossible so that no Abelian group can be our universal free group object.

Now we consider any group G with a generating set $\varphi(S)$. Any element x in G can be written:

$$x = \prod_{i=1,\ldots,N} \varphi(s_i)^{a_i}, s_i \in S, a_i \neq 0.$$

If there is a (non-trivial) relation $\mathscr{R}$ between the generators such that[89]:

$$\mathscr{R}: \qquad \prod_{i=1,\ldots,N} \varphi(s_i)^{a_i} = \prod_{j=1,\ldots,N'} \varphi(s_j)^{a'_j}, s_i, s'_i \in S, a_i, a'_i \neq 0.$$

Then if G is the universal object, this relation will be transported *to any other group* G':

$$\mathscr{R}': \qquad \prod_{i=1,\ldots,N} \psi(s_i)^{a_i} = \prod_{j=1,\ldots,N'} \psi(s_j)^{a'_j}, s_i, s'_i \in S, a_i, a'_i \neq 0.$$

As we can find groups where this latest relation will not be true, we conclude that any group where any non-trivial relation like $\mathscr{R}$ holds cannot be our universal object. We must then build a group *free* from any relations (e.g. the Free Group).

Solution:

We build the Free group with 2 generators: we consider two infinite cyclic groups G_a and G_b generated ,respectively, by a and b. We suppose that a and b are different.

Then we consider the sets of *words* $\mathbb{W}$ defined as succession of formal symbols $(x_1 x_2 \ldots x_n)$ (word of length n) and:

1) x_i must be either in C_a or in C_b,

2) x_i must not be a unit element of the group it belongs to,

3) and x_i, x_{i+1} must not belong to the same cyclic group.

---

[89]The relation $\mathscr{R}$ is equivalent to the fact that an element from G can be generated in two different ways from $\varphi(S)$

For example, the following words are elements from $\mathbb{W}$: (ab^3a^6b), (a), (aba), $(ab^2a^{-1}ba^{-12})$, etc... We provide this set of words with an operation $\star$ defined as follows:

- $\emptyset \star \emptyset = \emptyset$,

- $(x_1 x_2 \ldots x_n) \star \emptyset = \emptyset \star (x_1 x_2 \ldots x_n) = (x_1 x_2 \ldots x_n)$ (so that $\emptyset$ is the unit element).

If we suppose that $\star$ is defined for words of length $< n$, the we can define it for words of length n by the recursive formulas:

$(x_1 x_2 \ldots x_n) \star (y_1 y_2 \ldots y_n) =$

1. $(x_1 x_2 \ldots x_n y_1 y_2 \ldots y_n)$ if x_n and y_1 are not in the same group,

2. $(x_1 x_2 \ldots x_{n-1} z y_2 \ldots y_n)$ if x_n and y_1 are in the same group and $x_n \neq y_1^{-1}$, $z = (x_n \times y_1)$,

3. $(x_1 x_2 \ldots x_{n-1} y_2 \ldots y_n)$ if x_n and y_1 are in the same group and $x_n = y_1^{-1}$.

$(\mathbb{W}, \star)$ is a group, the inverse word of $(x_1 x_2 \ldots x_n)$ being given by $(x_n^{-1} x_{n-1}^{-1} \ldots x_1^{-1})$.

If we note $(a)^n$ the word $(a) \star (a) \star \ldots \star (a)$ (n times) we see easily that $(a)^n = (a^n)$ and so it becomes clear that every word $w \in \mathbb{W}$ can be written as

$$w = (x_1)^{\alpha_1} \star (x_2)^{\alpha_2} \star \ldots \star (x_N)^{\alpha_N}$$

with $x_i = a$ or b, $x_i \neq x_{i+1}$, $\alpha_i \neq 0$.

So that $(\mathbb{W}, \star)$ is a group with two generators: the word (a) and the word (b).

The group $G_2 = (\mathbb{W}, \star)$ is free from any possible non-trivial relations: indeed it is obvious that any word can be written in one and only one way because of the very nature of the elements of $\mathbb{W}$ that are free combination of symbols.

Finally we conclude that G_2 is the requested universal object .

Alternative solution

A proof of the general existence of Free Groups can be found in Algebra, S Lang and can be used as an alternative solution. We reproduce it here.

We want to prove the existence of a free group determined by S for any set S. For this we start to prove the following lemma:

There exists a set I and a family of groups G_i such that if $g:S \to G$ is a map of S into a group G, and that g generates G then G is isomorphic to some G_i .

If $card(S) < \infty$ then G is finite or enumerable.

If $card(S) = \infty$ then $card(G) \leq card(S)$ because G consists of finite products of $g(S)^i$.

Following the discussion mentioned above, if $card(S) < \infty$, we define a set T such as T infinite denumerable and if $card(S) = \infty$ then we define a set T such that $card(T) = card(S)$ so that in both case T has the maximal cardinality of finite products over $g(S)$.

For each non empty subset H of T, we define Γ_H as the set of group structures on H, that is to say the set of all groups that can be created by elements of H with an operation γ. For all $\gamma \in \Gamma_H$, $H_\gamma = (H, \gamma)$ is a group. Then if we consider the family:

$$\{H_\gamma\}_{\gamma \in \Gamma_H, H \subseteq T}.$$

This is the required family of groups [ii].

This proves the lemma.

For every $i \in I$, we note M_i the set of mappings of S into G_i. For each map $\varphi \in M_i$, we note $G_{i,\varphi}$ the set-theoretic product of G_i with the one element $\{\varphi\}$ so that $G_{i,\varphi}$ is the "same" group as G_i indexed by φ [iii].

We let:

$$F_0 = \prod_{i \in I} \prod_{\varphi \in M_i} G_{i,\varphi}$$

F_0 is the Cartesian product of the groups $G_{i,\varphi}$ and we define a map:

$$f_0 : S \to F_0$$

which sends S on the factor $G_{i,\varphi}$ by mean of φ itself[iv].

We contend that given a map $g: S \to G$ of S into a group G, there exists an homomorphism $\psi_\star: F_0 \to G$ making the usual diagram commutative:

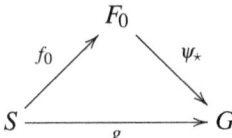

To prove this, we may assume that g generates G, simply by considering the subgroup of G generated by the image of g. By our lemma there exists $\lambda: G \to G_i$ making G isomorphic to G_i for an $i \in I$ and $\lambda \circ g$ is an element ψ of M_i.

We define $\pi_{i,\psi}$ as the projection in F_0 into the component $G_{i,\psi}$ and we put:

$$\psi_\star = \lambda^{-1} \circ \pi_{i,\psi}.$$

Then $\psi_\star$ is a group homomorphism and make the following diagram commutative:

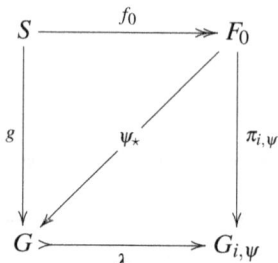

To end, we consider F as the subgroup of F_0 generated by $f_0(S)$.

We define f as f_0 considered as a map $S \to F$ ($f_0(S) \subseteq G$).

We let $g_\star$ be the restriction of $\psi_\star$ to F so that we have a unique homomorphism $g_\star \in \text{Mor}(F, G)$:

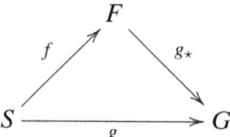

(F, f) is the Free group we wanted to build [v].

Notes

[i] *In the second case, G consists of finite products of a set $g(S)$ whose cardinality is obviously at most $card(S)$. So that, at most, G if formed of finite products over a set of infinite cardinality $card(S)$. Forming finite products over an infinite set does not change the cardinality, so that at most $card(G) = card(S)$.*

[ii] *It is not so evident that $\{H_\gamma\}_{\gamma,H}$ is the requested family. Let us consider an application g that generates a group G, then G is (at most) isomorphic to a set of finite products of elements of T, that is to say, a subset of T provided with a group structure. So that clearly, we can find $H \subseteq T$ and γ such that G is isomorph to H_γ.*

[iii] *That is to say, one just combine $\{\varphi\}$ to any elements of the group G_i.*

[iv] *f_0 is defined as follows: φ is in M_i so that φ determines the index i. So φ is an application $S \to G_{i,\varphi}$ ($G_{i,\varphi} = G_i$). We consider the "collective" action of all the possible φ's, they will send S to the components $G_{i,\varphi}$ of the Cartesian product. So that we might write, considering each φ as an application from S into F_0 that $f_0(S) = \prod_{i \in I} \prod_{\varphi \in M_i} \varphi(S)$.*

[v] *We summarize the process of building a free group:*

- *We build the Cartesian product F_0 of all the groups mapped from S by all possible applications from our Category $\mathfrak{C}_S$ and indexed by these applications*

- *We define an application $f_0 : S \to F_0$ by considering the product of all possible applications from our Category $\mathfrak{C}_S$, e.g. $f_0(x) = \prod_{i \in I} \prod_{\varphi \in M_i} \varphi(x)$*

- *We consider the restriction, f of f_0 to the subgroup G of F_0 generated by $f_0(S)$*

- *f defines the requested free group G, e.g. $G = < f_0(S) >$.*

52/828

Subject: Free Abelian group with two generators

Let $\mathfrak{AC}_2$ be the complete subcategory in $\mathfrak{C}_2$ whom objects are the Abelian groups with two distinguished generator. Describe the universal object in $\mathfrak{AG}_2$. This object is named *free Abelian group with two generators* .

SOLUTION:

Solution 1:

We consider the direct product $F_2 = C_a \times C_b$ where C_a and C_b are two cyclic infinite group with respective generators a and b, $a \neq b$ (and such that no relation exist between a and b).

Then F_2 is Abelian: if (x,y) and (x',y') are two elements of F_2 then $(x,y)(x',y') = (x',y')(x,y) = (xx',yy')$. Furthermore F_2 is generated by (a,e) and (e,b), e being the undifferentiated unit element in C_a and C_b so that $F_2 \in \mathfrak{AC}_2$.

Apart from the relation $\mathscr{R}$: $(a,e)(e,b) = (e,b)(a,e)$, there is no other relation in F_2. This relation $(\mathscr{R})$ will be transported by group homomorphisms to all other groups in the same subcategory $\mathfrak{AC}_2$, but this is is not a concern because all groups of $\mathfrak{AC}_2$ satisfy to $\mathscr{R}$. So that F_2 is our required universal object. If $G_2 = <u,v>$ is another element of the category $\mathfrak{AC}_2$, then the unique homomorphism $\varphi: F_2 \to G_2$ will be defined by:

$$\varphi(a,e) = u, \varphi(e,b) = v,$$
$$x \in F_2, x = (a^\alpha, b^\beta) \Rightarrow \varphi(x) = u^\alpha v^\beta.$$

We note that in the Abelian case, the construction of a free group is much simpler than in the general case. Obviously, the free group with n generators will be build as the direct product $C_{a_1} \times \ldots \times C_{a_n}$ where the C_{a_i}'s are infinite cyclic groups and where no non-trivial relation holds between the generators except commutativity[90].

Solution 2:

This solution deals with the general case so that no hypothesis are needed on the generating set S.

We recall that a group can be build from a free group and a set of relations in the following ways.

If F_0 is a free group with a generating set $f(S)$ and R is a subset of $f(S)$ then we define N as the smallest normal subgroup containing R. The quotient group F_0/N is then called *the group defined by the generators $f(S)$ and the relations R*. Indeed if φ is the canonical surjective homomorphism $G \to G/N$ then $R \subset \ker(\varphi)$ so that elements of G/N satisfy the relation $\varphi(R) = e$ (this is a non-trivial relation).

Thus, we may build a free Abelian group F_a from a free group F_0 by considering the relations $[x,y] = e, \forall (x,y) \in F_a$, and $[x,y]$ is the F_0 commutator operator $F_0 \times F_0 \to F_0$ defined by $[x,y] = xyx^{-1}y^{-1}$. If C is the commutator subgroup of F_0, that is to say the image of the commutator operator, then C is a normal subgroup in F_0 and we may define $F_a = F_0/C$. F_a is an Abelian group (because of the relations $[x,y] = e$ that now occur inside F_0/C [91]).

If ϕ is the group homomorphism from F_a to an other Abelian group G in the category $\mathfrak{C}_2$, then we derive an homomorphism ϕ' from F_a into G by noticing that: 1) $C \subseteq \ker(\phi)$ so that we can embed F_0/C into $F_0/\ker(\phi)$ by a canonical injection i. 2) $Im(\phi)$ is isomorphic to the quotient group $F_0/\ker(\phi)$ (a well-known property from basic group theory). We name j

---

[90]The Free Abelian Group with n generators is isomorph to $\mathbb{Z} \times \ldots \times \mathbb{Z}$ (the product being taken n times) or equivalently to $\bigoplus_{i=1}^{n}(\mathbb{Z})$

[91]Note that taking the quotient of a group by its commutator is a classical way to make it Abelian

this isomorphism $F_0/\ker(\phi) \to Im(\phi)$ and we see that the following diagram is commutative:

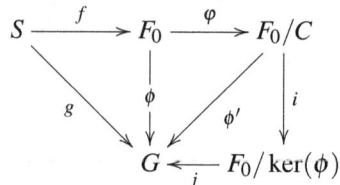

We note $k = \varphi \circ f$. If h is the application that maps S to F_0/C then $\varphi \circ f = h$, and φ is the unique morphism such that this property holds, because of the universal property of F_0. The same occurs for ϕ: it is the only morphism such that $\phi \circ f = g$. We have defined ϕ' as $j \circ i$ and we see that $\phi = j \circ i \circ \varphi$ so that $\phi' \circ \varphi = \phi$. This leads to $\phi' \circ \varphi \circ f = \phi \circ f$ or: $\phi' \circ (\varphi \circ f) = g$ or finally: $\phi' \circ k = g$.

Finally, in the subcategory $\mathfrak{AC}_2$ of $\mathfrak{C}_2$, made with the Abelian groups, we have build a universal object, $F_a = F_0/C$.

Notes

[i]The Free Group generated by a set Λ is isomorph to $\bigoplus_{\alpha \in \Lambda} \mathbb{Z}$.

[ii]We may define a Free Abelian Group G in terms of a *basis*: a set $X = \{c_\alpha\}_{\alpha \in A}$ is said to be a basis for an Abelian group if and only if for any element $x \in G$, x is expressed as a product of the generators in one and only one way: $x = \prod_{\alpha \in A} c_\alpha^{a_\alpha}$, $a_\alpha \in \mathbb{Z}$. Then a Free Abelian Group is simply an Abelian Group that has a basis and in this case, $G = \bigoplus_{\alpha \in A} \mathbb{Z}$.

53/828 ¶

Subject: Tensor Algebra

We consider $A_n(\mathbb{K})$ the category of associative algebras in a given field $\mathbb{K}$ with n distinguished generator. Describe the universal object in $A_n(\mathbb{K})$ (this object is named the *Tensor Algebra* on a vector space of dimension n over the field $\mathbb{K}$).

Extra Hint: Consider the algebra of polynomials with n non commutative variables $X_1, \ldots, X_n$ *over* $\mathbb{K}$.

SOLUTION:

The category $A_n(\mathbb{K})$ is made up with associative $\mathbb{K}$-Algebras $\mathbb{B}$ over a fixed field $\mathbb{K}$ with n generators. If X is a set such that $\#X = n$, these algebras are then described by an algebra $\mathbb{B}$ and an (injective) map φ from X to $\mathbb{B}$ such that $\mathbb{B}$ is generated by $\varphi(X)$, or equivalently that $\mathbb{B}$ is generated by $\{\varphi(a_i)\}_{i=1,\ldots,n}$, with $\{a_i\}_{i=1}^{i=n} = X$.

A morphism between two elements $(\mathbb{A}, \varphi)$ and $(\mathbb{B}, \psi)$ is an Algebra homomorphism $F: \mathbb{A} \rightarrow \mathbb{B}$ such that the following diagram is commutative:

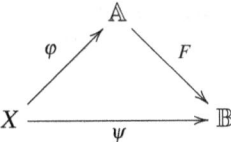

Discussion:

We refer to [?] for the definition of a $\mathbb{K}$-Algebra.

This problem is the building of a Tensor Algebra (or Free Algebra) $T(\mathbb{K}^n) = \bigoplus_{r=0}^{\infty} \otimes^r(\mathbb{K}^n)$, which suppose, from a theory point of view, the building of tensor product (Exercise 61). We give later an alternative solution that might be also of interest to the reader.

When trying to solve the problem, we might for example consider the case n=2, $\mathbb{K}=\mathbb{R}$: $\mathbb{C}$ is a $\mathbb{R}$-algebra with two generators: 1 and i so that $\mathbb{C} \in A_2(\mathbb{R})$. A universal object $\mathbb{A}$ must be linked to every other associative algebra $\mathbb{B}$ from $A_2(\mathbb{R})$ by an homomorphism F which send the generator to other generators. We let $e_i = \varphi(a_i)$ and $u_i = \psi(a_i)$. This homomorphism, if it exists is unique, because it is determined by:

1. $F(e_i) = u_i$, where e_i and u_i, $i = 1, \ldots, n$ are the respective generators of $\mathbb{A}$ and $\mathbb{B}$.

2. $F(\sum a_{i_1 \ldots i_p} e_{i_1} e_{i_2} \ldots e_{i_p}) = \sum a_{i_1 \ldots i_p} F(e_{i_1}) F(e_{i_2}) \ldots F(e_{i_p}), 1 \leq i_j \leq n$.

If $\mathbb{B}$ is an element of $A_2(\mathbb{R})$ with generators u and v, we may try $\mathbb{C}$ as a universal object and see what happens: then, if it exists the homomorphism F: $\mathbb{C} \rightarrow \mathbb{B}$ leads immediately to $F(i^2) = F(i)^2 = F(-1) = -F(1)$ so that we will transport the relation $u^2 = -v$ to the generators of $\mathbb{B}$. And since we can build an $\mathbb{R}$-algebra with 2 generators where this relation is not true, $\mathbb{C}$ cannot be the Free object we are looking for.

Reasoning in the same way as shown in exercise 51, we may try to build an algebra free of any kind of relation between the generators $\{e_i\}_{i=1,\ldots,n}$. In this purpose we look what happens when we generate the algebra from the generators: from the (non-commutative) product between the generators we get new elements of the form $e_{i_1} \times e_{i_2} \times \ldots \times e_{i_p}$ where p varies from 1 to ∞. If we do not want any non-trivial relation between the generators in our algebra, these elements must be all different so that we have to number them as follows:

$$e_{i_1} \times e_{i_2} \times \ldots \times e_{i_p} = e_{i_1 i_2 \ldots i_p}.$$

Following this, we will have for two elements $e_{i_1 i_2 \ldots i_p}$ and $e_{j_1 j_2 \ldots j_q}$ the rule:

$$e_{i_1 i_2 \ldots i_p} \times e_{j_1 j_2 \ldots j_p} = e_{i_1 i_2 \ldots i_p j_1 j_2 \ldots j_p}$$

and we have to claim that all these elements are different. At this stage, we can consider our object as a vector space over $\mathbb{K}$ provided with a basis made of all the $e_{i_1 i_2 \ldots i_p}$!

Solution:

We consider set S_n of sequences $\{i_1, i_2, \ldots, i_p\}, 1 \le p \le n$. For two elements I and I' of S_n we note II' the element of S_n created by the concatenation of I and I'. We define a vector space $T(n)$ provided with an infinite basis $\{e_I\}_{I \in S_n}$ and a multiplication noted $\otimes$ defined by:

$$e_I \otimes e_{I'} = e_{II'}.$$

This operation send elements of the base into itself so that we can use it to extend the operation $\otimes$ over any element of $T(n)$ by:

$$\left(\sum a_I e_I\right) \otimes \left(\sum b_J e_J\right) = \sum a_I b_J e_I \otimes e_J = \sum a_I b_J e_{IJ}$$

where the sums are taken over a finite sequence of elements in I and J.

$T(n)$ provided with $\otimes$ is an associative algebra and is generated by the elements $e_1, e_2, \ldots e_n$. Indeed it is clear that every $x \in T(n)$ can be written $x = \sum_{I \in S_n} a_I e_I$, $a_I \in \mathbb{K}$ or:

$$x = \sum_{I \in S_n} a_{i_1 i_2 \ldots i_p} e_{i_1 i_2 \ldots i_p}, \quad a_{i_1 i_2 \ldots i_p} \in \mathbb{K}$$

and we can also read this as:
$x = \sum_{I \in S_n} a_{i_1 i_2 \ldots i_p} e_{i_1} \otimes e_{i_2} \otimes \ldots \otimes e_{i_p}$, $a_{i_1 i_2 \ldots i_p} \in \mathbb{K}$.

This means that every element in $T(n)$ can be written as the sum of products of $e_i's$, $i = 1, \ldots, n$.

Besides, the product of two elements $x = \sum_{I \in S_n} a_I e_I$, $a_I \in \mathbb{K}$ and $y = \sum_{J \in S_n} b_J e_J$, $b_J \in \mathbb{K}$ is again in $T(n)$ because it is written:

$$
\begin{aligned}
xy &= \left(\sum_{I \in S_n} a_I e_I \right)\left(\sum_{J \in S_n} b_J e_J \right), \\
&= \sum_{I,J \in S_n} a_I b_J e_{IJ}.
\end{aligned}
$$

The other properties that make up $T(n)$ an associative algebra are easily verified.

Repeating ourselves, there cannot exists any non-trivial algebraic relation[92] between the generators in $T(n)$ (such as for example $i^2 = -1$ in $\mathbb{C}$): If we could find such a relation, that would mean a non-trivial linear dependence of the form $\sum_i a_i e_{I_i} = 0$ which is not possible because, by construction, $\{e_I\}$ is a basis of $T(n)$.

Now we must prove that for all associative Algebra there exist an homomorphism F that makes the following diagram commutative (i is the application $X \to T(n)$, $i(a_i) = e_i, 1 \le i \le n$):

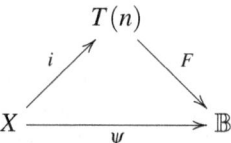

We can first check that F exists because there is no non-trivial relation in $T(n)$ so that the definition of the homomorphism given above is not contradictory. e-g there is a unique representation of any element $x \in T(n)$ by: $x = \sum a_I e_I$ and then F is defined without ambiguity by $F(x) = \sum a_I F(e_I)$, $F(e_I)$ being given by $F(e_{i_1} \ldots e_{i_r}) = F(e_{i_1}) \ldots F(e_{i_r})$, we can assume that F maps e_i to the the generators of $\mathbb{B}$ so that our diagram is commutative.

---

[92]E.g. a relation between linear combinations of products of generators

Another alternative way to see this, is to use the universality of the tensor product $E_1 \otimes E_2$ of two $\mathbb{K}$-vector spaces (or more generally of two modules over a ring) as defined in Exercise 61.

First we define the *r*th tensor product $E_1 \otimes E_2 \otimes \ldots \otimes E_r$ by the following recursive formula:

$$E_1 \otimes E_2 \otimes \ldots \otimes E_{r+1} = (E_1 \otimes E_2 \otimes \ldots \otimes E_r) \otimes E_{r+1}.$$

We can check at once that $(E_1 \otimes E_2 \otimes \ldots \otimes E_r) \otimes E_{r+1}$ is equivalent to $E_1 \otimes (E_2 \otimes \ldots \otimes E_{r+1})$ so that the tensor product of r vector spaces is well defined.

In our case, we can define for $r > 0$, the *r*th tensor power of E, $\otimes^r E$, by:

$$\otimes^{r+1} E = (\otimes^r E) \otimes E = E \otimes (\otimes^r E)$$

and we can define that the 0th-tensor power of E to be the unit element of the field $\mathbb{K}$, if E is a $\mathbb{K}$ vector space.

We can define for $r \geq 0$, the application g_r:

$$E \times \ldots \times E \to \mathbb{B},$$
$$(x_1, \ldots, x_r) \to g_r(x_1, \ldots, x_r) = g(x_1) \ldots g(x_r).$$

If we note f_r the application:

$$E \times \ldots \times E \to \otimes^r E,$$
$$(x_1, \ldots, x_r) \to f_r(x_1, \ldots, x_r) = x_1 \otimes \ldots \otimes x_r$$

then we have the following commutative diagram:

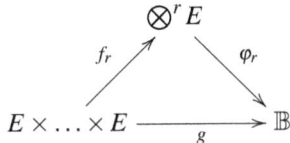

With φ_r uniquely defined by $\varphi_r(x_1 \otimes \ldots \otimes x_r) = g(x_1) \ldots g(x_r)$.

If we define the direct sum $T(E) = \bigoplus_{r=0}^{r=\infty} \otimes^r E$, then we may define the application φ by:

$$T(E) \to \mathbb{B},$$

$$\sum_{r=0}^{\infty} \lambda_{i_1} \ldots \lambda_{i_r} e_{i_1} \otimes \ldots \otimes e_{i_r} \to \sum_{r=0}^{\infty} \lambda_{i_1} \ldots \lambda_{i_r} g(e_{i_1} \otimes g(e_{i_r}).$$

We see that φ is an algebra homomorphism, if we define f as the injection $X \to T(E)$, then we see that φ is such that $\varphi \circ f = g$ and that φ is unique. Of course, we can easily see now that the operation $\otimes$ we have defined at the beginning of the solution is the tensor product as defined in Exercise 61.

One other way to define the homomorphism φ from T(E) to any other algebra $\mathbb{B}$ is to note that $T(E)$ is the direct limit of the directed family $\{\otimes^r(E)\}_{r \in \mathbb{N}}$: e.g. $T(E) = \varinjlim \otimes^r(E)$ (see Exercise 63), and by definition, we get the following commutative diagram[93]:

---

[93]Where the applications f_r, f_{rs} and g_r, $(r,s) \in \mathbb{N}^2$ may be deduced from what we said previously

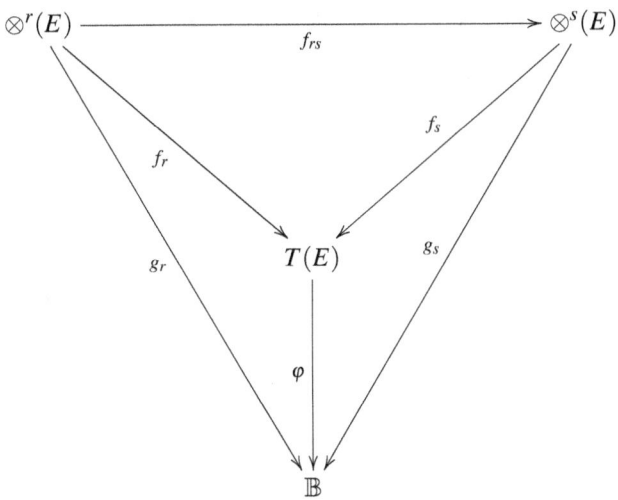

Alternative solution:

We consider the algebra $\mathbb{A} = \mathbb{K}^{nc}[X_1, \ldots, X_n]$ of non-commutative poly-nomials[94] with n variables. That is the say the algebra made with polynomials

$$P(X_1, X_2, \ldots, X_n) = \sum_{1 \leq i_j \leq n} a_{i_1 i_2 \ldots i_r} X_{i_1} X_{i_2} \ldots X_{i_r},$$

(The sum being finite).

$\mathbb{A}$ is naturally generated by the variables $X_1, X_2, \ldots, X_n$ and is associative. The family $\{X_{i_1} X_{i_2} \ldots X_{i_r}\}_{1 \leq i_j \leq n, j \leq r, r \geq 0}$ form a basis of $\mathbb{A}$ over $\mathbb{K}$.

We can then define without ambiguity an homomorphism F from $\mathbb{A}$ to any other algebra $\mathbb{B}$ with generators $B_1, \ldots, B_n$ by

$$F(X_i) = B_i,$$

---

[94]E.g. $X_i X_i \neq X_j X_i, i \neq j$

$$F\left(\sum_{1\leq i_j\leq n} a_{i_1 i_2 \ldots i_r} X_{i_1} X_{i_2} \ldots X_{i_r}\right) = \sum_{1\leq i_j\leq n} a_{i_1 i_2 \ldots i_r} F(X_{i_1}) F(X_{i_2}) \ldots F(X_{i_r}).$$

Because there may be only one universal object, $\mathbb{K}^{nc}[X_1,\ldots,X_n]$ and $T(\mathbb{K}^n)$ must be isomorphic. However, we can easily see that the tensor algebra and the non commutative algebra are equivalent *in the context of this exercise* .

54/828 ¶

Subject: Universal object in $CA_n(\mathbb{K})$

Show the existence of a Universal (repelling) Object in the subcategory $CA_n(\mathbb{K})$ of $A_n(\mathbb{K})$ made with the commutative algebras in $A_n(\mathbb{K})$.

SOLUTION:

A well-known method to transform an algebra A into a commutative algebra A' is to consider the quotient $A' = A/C(A)$ where $C(A)$ is the bilateral ideal defined by $\{[x,y] = xy - yx, (x,y) \in A\}$ ($[x,y]$ is the bracket product). Indeed, A' is commutative because $\overline{[x,y]} = \bar{0}$ that is to say $\bar{x}\bar{y} - \bar{y}\bar{x} = \bar{0}$. So if we consider $T_n(\mathbb{K})$ the Tensor algebra with n generators, $T_n(\mathbb{K})$ is the universal object of the category $A_n(\mathbb{K})$.

The reasoning is exactly the same as for Exercise 52 (e.g. building a free universal object in the subcategory of Abelian groups with n generators) except that the free algebra $T_n(\mathbb{K})$ stands for the free group F_0, so that we will consider the quotient $T_n(\mathbb{K})/C(T_n(\mathbb{K}))$ and build the following commutative diagram[95]:

---

[95] See Exercise 52 for the definition of the mappings and proof that the diagram is commutative

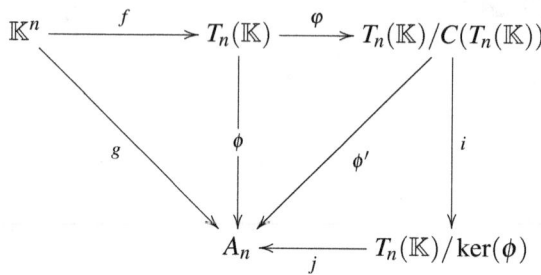

$T_n(\mathbb{K})/C(T_n(\mathbb{K}))$ is the requested universal object in $CA_n(\mathbb{K})$.

55/828 ¶

Subject: Free Lie Algebra

Show the existence of a universal object in the Category $LA_n(\mathbb{K})$ of Lie Algebras, over the field $\mathbb{K}$, with n generators.

SOLUTION:

We recall that a non associative algebra $\mathbb{A}$ is said to be a Lie algebra if its multiplication law satisfies the Lie conditions:

$$x^2 = 0, (xy)z + (yz)x + (zx)y = 0. (\forall(x,y,z) \in \mathbb{A}^3).$$

The second condition is called the Jacobi identity. First, we must define the category $LA(X,\mathbb{K})$ of Lie Algebras over a field $\mathbb{K}$ generated by a set X. We define it as the objects $(\mathfrak{g}, f)$, $\mathfrak{g}$ being a Lie algebra and f, a mapping from X to $\mathfrak{g}$, if χ is a Lie-homomorphism between $\mathfrak{g}$ and $\mathfrak{g}'$ such that $f' = \chi \circ f$, then χ is a morphism from the object $(\mathfrak{g}, f)$ to the object $(\mathfrak{g}', f')$:

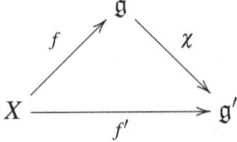

Indeed, if $(\mathfrak{g}, f)$, $(\mathfrak{g}', f')$ and $(\mathfrak{g}'', f'')$ are three objects in the category, then for χ, a Lie Homomorphism $\mathfrak{g} \to \mathfrak{g}'$ and χ', a Lie-homomorphism $\mathfrak{g}' \to \mathfrak{g}''$, we have the following diagram:

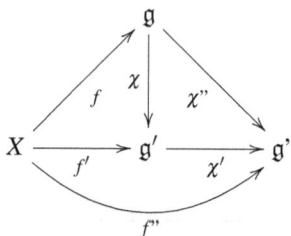

Then if f" is defined by $f" = \chi' \circ f' = \chi' \circ \chi \circ f$ so that $f" = (\chi' \circ \chi) \circ f$ and therefore, $\chi" = \chi' \circ \chi$ is a Lie-Homomorphism from $\mathfrak{g} \to \mathfrak{g}"$ such that $f" = \chi" \circ f$ and then $\chi" \in Mor((\mathfrak{g},f),(\mathfrak{g}",f"))$, so that the morphisms of the category are well-defined and the Free Lie group generated by the set X will be defined as the universal object in this category. We consider here that X is finite with cardinality $= n$.

Solution:

We follow the hint and we build the sets $\{E_n\}_{n=1}^{n=\infty}$ defined by: $E_1 = \{e_1,\ldots,e_n\},\ldots$, $E_n = \bigsqcup_{k+l=n} E_k \times E_l$, $n \geq 2$, and we build the set $M = \bigsqcup_{n=1}^{n=\infty} E_n$, we provide M with the law $x \times y = (x,y)$, then we consider $K(M)$, the algebra generated by M.

At first, we could think that $K(M)$ is the tensor Algebra defined in Exercise 53, but it is *not*; it is a more general algebra, named the *Free Magma Algebra with n generators*[96], it is defined to be the more general free non-associative Algebra with n generators at the difference with the Tensor Algebra which is associative.

For example: $E_2 = E_1 \times E_1 = \{(e_i,e_j), 1 \leq (i,j) \leq n\}$, $E_3 = (E_2 \times E_1) \sqcup (E_1 \times E_2) = \{((e_{i_1},e_{i_2}),e_{i_3}), i_1,i_2,i_3 \leq n\} \sqcup \{(e_{i_1},(e_{i_2},e_{i_3})), i_1,i_2,i_3 \leq n\}$ and $E_3 \neq E_1^3$ since, in general, $((e_{i_1},e_{i_2}),e_{i_3}) \neq (e_{i_1},e_{i_2},e_{i_3})$.

The basis we are building on is in fact the more general Free non-associative basis for an Algebra that we can be found.

We can represent the elements of M by the symbolic expressions:

---

[96]M is named the *Free Magma*

$$\diamond_{i_0} e_{i_1} \diamond_{i_1} e_{i_2} \ldots \diamond_{i_{k-1}} e_{i_k} \diamond_{i_k}$$

where the symbols $\diamond_{i_j}$ are some special combinations of open and closing parenthesis and commas.

The operation on the Free Magma Algebra is then defined by:

$$\diamond_{i_0} e_{i_1} \diamond_{i_1} e_{i_2} \ldots \diamond_{i_{k-1}} e_{i_k} \diamond_{i_k} \star \diamond_{j_0} e_{j_1} \diamond_{j_1} e_{j_2} \ldots \diamond_{i_{l-1}} e_{i_l} \diamond_{i_l}$$

$$=$$

$$(\diamond_{i_0} e_{i_1} \diamond_{i_1} e_{i_2} \ldots \diamond_{i_{k-1}} e_{i_k} \diamond_{i_k}, \diamond_{j_0} e_{j_1} \diamond_{j_1} e_{j_2} \ldots \diamond_{i_{l-1}} e_{i_l} \diamond_{i_l}).$$

For example:

$$(((e_1, e_2), e_3)) \star (e_6, e_7) = ((((e_1, e_2), e_3)), (e_6, e_7)).$$

We can check at once that K(M) is an algebra. K(M) (that we can note also K(X)) is much bigger than T(X) (here $T(\mathbb{K}^n)$), the tensor algebra created from X.

We form next the two-sided ideal I generated by the elements $Q(a) = \{a.a, a \in K(X)\}$ and $J(a,b,c) = \{a(bc) + b(ac) + c(ab), (a,b,c) \in K(X)\}$.

Of course, our goal is to show that $L(X) = K(X)/I$ will define the Free Lie Algebra as required[97].

Two applications $\phi : X \to L(X)$ and $f : X \to \mathfrak{g}$ can be extended, respectively, to the homomorphisms $\tilde{\phi} : K(X) \to L(X)$ and $\tilde{f} : K(X) \to \mathfrak{g}$ by making:

$$\tilde{\phi}(\diamond_{i_0} e_{i_1} \diamond_{i_1} e_{i_2} \ldots \diamond_{i_{k-1}} e_{i_k} \diamond_{i_k}) = \phi(e_{i_1}) \ldots \phi(e_{i_k})$$

and:

$$\tilde{f}(\diamond_{i_0} e_{i_1} \diamond_{i_1} e_{i_2} \ldots \diamond_{i_{k-1}} e_{i_k} \diamond_{i_k}) = f(e_{i_1}) \ldots f(e_{i_k}).$$

These extensions will be unique since both function and extension will coincide on the generators.

---

[97]The quotient of an Algebra A by an ideal I is defined as the quotient of A by I from additive group point of view

In fact $\tilde{\phi}$ is the canonical mapping $K(X) \to K(X)/I$.

We see that $\tilde{f}(Q(a)) = 0$ and $\tilde{f}(J(a,b,c)) = 0$ because of the Jacoby identity in Lie Algebras.

Additionally, $L(X)$ is a Lie algebra because for all elements $\bar{a}, \bar{b}, \bar{c}$ in $L(X)$, we have that $a(bc) + b(ca) + c(ab) = \bar{0}$ and $\overline{a.a} = 0$.

So that I is normal in $\ker(\tilde{f})$ and then we can factor $\tilde{f}$ by $\tilde{\phi}$ in $\tilde{f} = F \circ \tilde{\phi}$ [98].

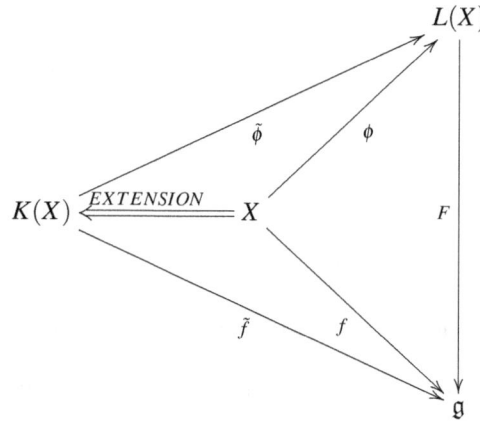

$M(X)$ =Free Magma,
$K(X)$ = Free Magma Algebra,
$L(X) = K(X)/I$.
We see immediately that :

$$\tilde{f}|_X = F \circ \tilde{\phi}|_X,$$
$$\text{e.g. } f = F \circ \phi.$$

The unicity of F comes from the unicity of the extensions $\tilde{\phi}$ and $\tilde{f}$.

Thus $L(X)$ is the required object.

---

[98] $Im(\tilde{f}) \approx K(X)/\ker(\tilde{f})$, $Im(\tilde{\phi}) = L(X) \approx K(X)/\ker(\tilde{\phi}) \Rightarrow \ker(\tilde{\phi}) \triangleleft \ker(\tilde{f})$ and this implies that we can find an homomorphism $F : K(X)/\ker(\tilde{\phi}) \to K(X)/\ker(\tilde{f})$

Notes

[i]We note, according to the hint, that the idea of taking the quotient space A/I where I is a two sided ideal (or a group if A is a group) generated by elements of A spanned by some algebraic functions $R(a_1, \ldots, a_N)$, will ensure us that $\forall \overline{a}_i'$ in A/I the following algebraic relations[99] holds:

$$\overline{R(a_1, \ldots, a_N)} = R(\overline{a}_1, \ldots, \overline{a}_N) = 0.$$

We have used successfully this process in some of the previous exercises in order to build (free) universal objects.

[ii]The Tensor Algebra can itself be defined from the Magma Algebra by taking the quotient $K(X)/S$ where S is the two sided ideal generated by elements $S(a,b,c) = \{(ab)c - a(bc), (a,b,c) \in K(X)\}$ so that the Tensor Algebra is directly obtained from the Free Magma Algebra by getting rid of the associativity (this is equivalent to the use of a forgetful functor). We note then that the free objects we have defined in Exercises 53,54 can also be expressed in term of the quotient of the free magma algebra.

[iii]We mention here an other possibility to build our Free object from the Free Magma by using the *P. Hall Sets* from the Free Magma.

If we define the length l of an element of the Free Magma Algebra by $l(\diamond_{i_0} e_{i_1} \diamond_{i_1} e_{i_2} \ldots \diamond_{i_{k-1}} e_{i_k} \diamond_{i_k}) = i_k$, then P. Hall Sets are a subset of $K(X)$ such that :

1) P. Hall Sets are totally ordered,

2) $X \subset H$,

3) if $(u,v) \in H$ and $l(u) < l(v)$ then $u < v$,

4) for each $u \in M - X$, we let $u = vw$ the unique decomposition of u with $(v,w) \in K(X)$[100].Then $u \in H$ if and only if :

4a) $v \in H$, $w \in H$ and $v < w$,

4b) either $w \in X$ or $w = w'w''$ with $w' \in H$,$w'' \in H$ and $w' \leq v$.

The images of P. Hall Sets in $L(X)$ are named P. Hall Basis of the Free Lie Algebra. The proof that the P. Hall basis are actually a basis of $L(X)$ may be found in [?].

Then if $H(X)$ is the algebra generated by H, we have $H(X) = L(X)$. L(X) is then a graded Algebra (like the tensor algebra or the magma algebra) and $L(X) = \bigoplus_{n=1}^{n=\infty} L_n(X)$, where $L_n(X)$ is the algebra generated by the Hall elements of length n (e.g. by the $\diamond_{i_0} e_{i_1} \diamond_{i_1} e_{i_2} \ldots \diamond_{i_{n-1}} e_{i_n}$'s).

Among the interesting properties of L(X), we can show that the amount of Hall elements in $L_n(X)$ (e.g. $dim(L_n(X))$) is obtained by the formula[101] :

$$dim L_n(X) = \frac{1}{n} \sum_{d|n} \mu(d)(card(X))^{n/d}$$

μ being the Möbius function in $(\mathbb{N}, |)$ (see Exercise 10b).

---

[99]Provided they are well-defined in the algebraic structure we are working with

[100]Any element u of $M - X$ has a unique decomposition $u = vw, (v,w) \in K(X)$, this can be easily checked

[101]See, again, [?] for a proof of this result

^{iv} In fact, we see that $T_n(\mathbb{K})^{102}$ can be turned into a Lie Algebra T_L by making $[x,y] = x \otimes y - y \otimes x$ (this is valid for any associative algebra $\mathbb{A}$, which can be turned into a Lie associative algebra A_L by using the product $x \star y = [x,y] = xy - yx$) — so that we can ask ourselves why $L(X)$ is not simply $T_L(X)$ for T is the universal object in the category of associative algebras... The answer is that the Lie algebras are not necessarily associative and hence they do not form a subcategory of $CA_n(X)$. But in fact we can show that, instead, $L(X)$ is the subalgebra of T_L generated by X. Let us say that $X = \{x_1, \ldots, x_n\}$. If we consider the set of elements $\{[\ldots [x_{i_1}, x_{i_2}], x_{i_3}], \ldots, x_{i_m}], 1 \le i_j \le n, 1 \le j \le m, m \in \mathbb{N}\}$, then this set will span a linear subspace $F(X)$ of T_L, from Jacobi's relations. Since $F(X)$ contains the x_i's, $F(X)$ will be the Free Lie Algebra. This gives us an other way to build the Free Algebra[103].

---

[102] Since we have introduced the notation $K(X)$, we could also use the notation $T(X)$ for the tensor algebra but this is not standard and we will note it T in what follows

[103] Unfortunately, the set we have considered is *not* a basis and generates many redundant elements

56/828 ¶

Subject: Enveloping Algebra

Let $\mathfrak{S}$ be a Lie algebra over the field K of characteristic 0. We consider the category $K(\mathfrak{S})$ whose objects are the linear mappings φ from $\mathfrak{S}$ into the associative algebras (this category is dependent on $\mathfrak{S}$) with the following property:

$$\varphi([x,y]) = \varphi(x)\varphi(y) - \varphi(y)\varphi(x).$$

We define a morphism from an object $\varphi : \mathfrak{S} \to \mathbb{A}$ into an object $\psi : \mathfrak{S} \to \mathbb{B}$, an homomorphism $\xi : \mathbb{A} \to \mathbb{B}$ such that the following diagram:

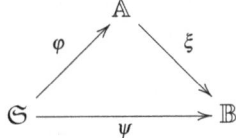

be commutative.
Show that $K(\mathfrak{S})$ have a universal object

$$\varphi_0 : \mathfrak{S} \to U(\mathfrak{S}).$$

The Algebra $U(\mathfrak{S})$ is named the (Universal) Enveloping Algebra[104] or Associative Hull for $\mathfrak{S}$.

SOLUTION:

---

[104]We will note it the U.E.A of $\mathfrak{S}$

We first note that, in the definition of the previous category, φ and ψ are in fact homomorphisms between $\mathfrak{S}$ and respectively $\mathbb{A}_L$ and $\mathbb{B}_L$[105], so that we can consider instead the following diagram where φ and ψ are (Lie-) homomorphisms:

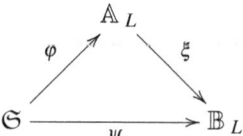

We first define $T(\mathfrak{S})$ the tensor algebra of $\mathfrak{S}$, we define it as the tensor algebra of $\mathfrak{S}$ seen as a vector space. If $\mathfrak{S}$ is a Lie algebra over the field $\mathbb{K}$ with n generators then $T(\mathfrak{S}) = T(\mathbb{K}^n)$[106]. The objects in $K(\mathfrak{S})$ can be represented by pairs $(\mathbb{A}, \varphi)$, φ being a linear mapping $\mathfrak{S} \to \mathbb{A}$, $\mathbb{A}$ being an associative algebra. A morphism between two objects $(\mathbb{A}, \varphi)$ and $(\mathbb{B}, \phi)$ is an algebra-homomorphism χ such that $\chi \circ \varphi = \phi$.

Again, it would be tempting to claim that $T(\mathfrak{S})$ is our universal object in $K(\mathfrak{S})$ but the morphisms in this category are Lie-homomorphisms, not general homomorphisms $\mathfrak{S} \to \mathbb{A}$. Then we need to force the relations $R(x,y) = 0$ where $R(x,y) = [x,y] - (x \otimes y - y \otimes x)$ in $T(\mathfrak{S})$ so that homomorphisms $T(\mathfrak{S}) \to \mathbb{A}$ will be automatically turned into Lie-homomorphisms: For this we consider the two-sided ideal J spanned by elements $J(x,y) = \{[x,y] - (x \otimes y - y \otimes x), (x,y) \in \mathfrak{S}\}$. We have $J \subset \mathfrak{S} \subset T(\mathfrak{S})$. Next we consider $U = T(\mathfrak{S})/J$ and we want to show that $U = U(\mathfrak{S})$ is the U.E.A we try to find.

We have the following situation:

---

[105]If $\mathbb{A}$ is an associative algebra, we define $\mathbb{A}_L$ to be the Lie algebra equipped with the product $[x,y] = xy - yx$

[106]So that we could note it $T(X)$

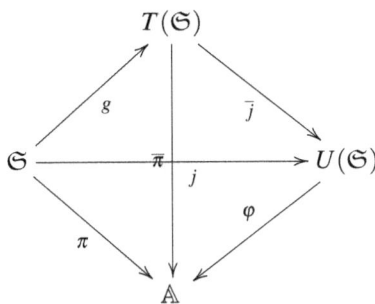

We are being given j and π two homomorphisms from, respectively, $\mathfrak{S} \to U(\mathfrak{S})$ and $\mathfrak{S} \to \mathbb{A}$. J can be extended to $\bar{j} : T(\mathfrak{S}) \to U(\mathfrak{S})$ and π can be extended to: $\bar{\pi} : T(\mathfrak{S}) \to \mathbb{A}$. We also can find $g : \mathfrak{S} \to T(\mathfrak{S})$ such that: $\pi = \bar{\pi} \circ g$ and $j = \bar{j} \circ g$.

Furthermore $\bar{j}$ is the canonical mapping $T(\mathfrak{S}) \to T(\mathfrak{S})/J = U(\mathfrak{S})$.

Indeed $(T(\mathfrak{S}), g)$ has the universal property in the category of associative algebras so that there is a unique mapping $\bar{\pi}$ between $(T(\mathfrak{S}), g)$ and $(\mathbb{A}, \pi)$ and there is also for the same reasons a unique morphism $\bar{j}$ mapping $(T(\mathfrak{S}), g)$ and $(U(\mathfrak{S}), j)$.

The canonical mapping $T(\mathfrak{S}) \to T(\mathfrak{S})/J$ is also a morphism from $(T(\mathfrak{S}), g)$ to $(U(\mathfrak{S}), j)$ so that $\bar{j}$ must be this mapping.

The key point is now to note that $\bar{\pi}([x,y] - (x \otimes y - y \otimes x))$ must be $= 0$ because $\bar{\pi}$ is a Lie-homomorphism and thus $\bar{\pi}([x,y]) = \bar{\pi}(x)\bar{\pi}(y) - \bar{\pi}(y)\bar{\pi}(x)$. So that $\ker(\bar{j})$ is normal in $\ker(\bar{\pi})$ and, following the notes of Exercise 56, we see that $\bar{\pi}$ can be factored by $\bar{j}$, let us say: $\bar{\pi} = \varphi \circ \bar{j}$.

Then we have found an homomorphism $\varphi : U(\mathfrak{S}) \to \mathbb{A}$. Let us check that φ makes our diagram commutative:

$$\begin{aligned}
\varphi \circ j &= \varphi \circ (\bar{j} \circ g), \\
&= (\varphi \circ \bar{j}) \circ g, \\
&= \bar{\pi} \circ g, \\
&= \pi.
\end{aligned}$$

So that φ is unique and well-defined as a morphism: Then $U(\mathfrak{S}) = T(\mathfrak{S})/J$ is the U.E.A we were looking for[107].

Notes

[i]One way to build the U.E.A, $U(\mathfrak{S})$, in the case where $\mathfrak{S}$ has n generators is to use the Poincare-Birkhoff-Witt theorem (PBW Theorem) which claims that the monomials:

$$\{j(e_{i_1})^{\alpha_{i_1}} \ldots j(e_{i_j})^{\alpha_{i_j}}\}_{1 \leq i_1 < i_2 < \ldots < i_j \leq n}$$

forms a basis of $U(\mathfrak{S})$.

---

[107]For more details, see [?]

57/828 ¶

Subject: Associative Hull of a Lie algebra with n generators

Show that the Associative Hull (see Exercise 56) of a Lie Algebra with n generators is isomorph to a Tensor Algebra over a space of dimension n.

SOLUTION:

We want to show that when $\mathfrak{S}$ is a Free Lie Algebra $\mathfrak{S} = F(X)$ then $U(\mathfrak{S}) = T(E)$, the tensor algebra of E, where E is the vector space generated by X (in the context of the exercise, X is finite and $E = \mathbb{K}^n$).

There are several ways to prove this result:

Following the hint we will first use only Category theory. When $\mathbb{A}$ is an associative algebra, we will note $\mathbb{A}_L$ the associative Lie Algebra consisting in $\mathbb{A}$ provided with the product $x \star y = [x, y] = xy - yx$.

The idea is to use the universality of F(X) in the category of Lie algebras and either start from the universality $U(\mathfrak{S})$ to show that it is also universal in the category of associative algebras or to start from the universality of $T(E)$ and show it is universal in the category $K(\mathfrak{S}) = K(F(E))$ of Lie-homomorphisms from $\mathfrak{S}$ to associative algebras.

Solution 1) $U(\mathfrak{S}) = T(E)$.

We want to show that $U(\mathfrak{S})$ is universal in the category of associative algebras over a field $\mathbb{K}$ and generated by a set X. Let us suppose we have the following diagram in the category of associative algebras:

TARGET:

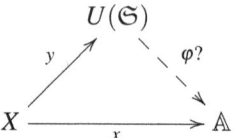

We need to show that for any object $(\mathbb{A}, x)$, we can find a unique morphism φ such that this diagram will be commutative (we also need to define y).

We know that, in the category of Lie algebra we have the following situation (and we can define precisely y there), $F(E)$ being the universal object:

CATEGORY OF LIE ALGEBRAS:

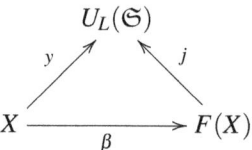

$U(\mathfrak{S})$ being considered here as an associative Lie algebra (e.g. $U_L(\mathfrak{S}) \to U(\mathfrak{S})$, where $U_L(\mathfrak{S})$ is $U(\mathfrak{S})$ provided by the bracket product $x \star y = [x,y]_{U(\mathfrak{S})} = xy - yx$).

We use again the universal property of F(X) in order to factor x. Indeed if we consider $\mathbb{A}$ as an associative Lie algebra (making $\mathbb{A}_L \to \mathbb{A}$) we have the following commutative diagram in the category of Lie groups:

CATEGORY OF LIE ALGEBRAS:

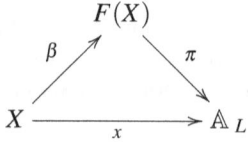

From the mapping $j : F(X) \to U(\mathfrak{S})$ we can now use the universal property of $U(\mathfrak{S})$, e.g. to be the U.E.A of $F(X)$, so that we have the following diagram:

CATEGORY OF LIE-HOMOMORPHISMS FROM LIE ALGEBRAS TO ASSOCIATIVE ALGEBRAS:

$$U_L(\mathfrak{S})$$

j φ

$$\mathfrak{S} = F(X) \xrightarrow{\quad \pi \quad} \mathbb{A}_L$$

and φ is a unique homomorphism. j being a Lie homomorphism, so will be φ.

But if we group all this, we have the following diagram:

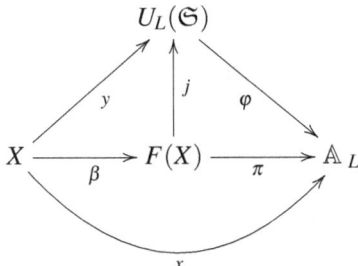

Here, we note that φ is an homomorphism from $U(\mathfrak{S})$ into $\mathbb{A}$ and that x and y are the same applications whatever we consider U or U_L and $\mathbb{A}$ or $\mathbb{A}_L$.

This diagram is commutative, indeed:

$\pi = \varphi \circ j$ so that:

$$
\begin{aligned}
\pi \circ \beta &= \varphi \circ j \circ \beta, \\
&= \varphi \circ (j \circ \beta), \\
&= \varphi \circ y.
\end{aligned}
$$

Finally we have:$x = \pi \circ \beta = \varphi \circ y$.

Finally this shows that $U(\mathfrak{S})$ is the tensor algebra $T(\mathfrak{S})\ (= T(E)\ {}^{108})$

Solution 2) $T(E) = U(\mathfrak{S})$

We start from $T(E)$ the tensor algebra of E and we want to find $\bar{\pi}$, a morphism $T(E) \to \mathbb{A}$, for any object $(\mathbb{A}, \pi)$ in the category $K(\mathfrak{S})$ of Lie-homomorphisms $\mathfrak{S} = F(E) \to \mathbb{A}$.

TARGET:

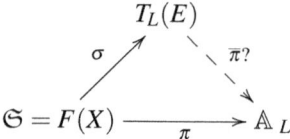

and σ is defined by the following diagram in the category of Lie groups:

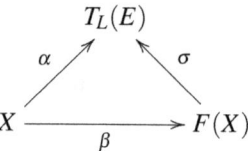

If we *define* x by $x = \pi \circ \beta$ then we have the following situation in the category of associative algebras:

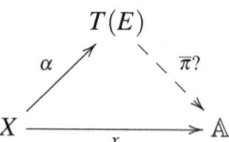

and of course as $T(E)$ is the universal object of this category, we can define uniquely $\bar{\pi}$ as the homomorphism $T(E) \to \mathbb{A}$ that makes the diagram

---

[108] If F is a Lie algebra generated by X, then $T(F)$ is the tensor algebra of a vector space generated by X because when building $T(F)$ we forget the structure of Lie Algebra only to keep the vector space structure.

commutative. We can obviously extend $\overline{\pi}$ to an homomorphism from $T_L(E)$ into $\mathbb{A}_L$.

Grouping all this, we have:

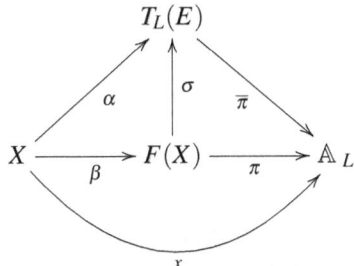

We need now to check that the mapping $\overline{\pi}$ we have defined will make our diagram commutative (in the category $K(\mathfrak{S}) = K(F(X))$).

$(\overline{\pi} \circ \sigma) \circ \beta = \overline{\pi} \circ \alpha$ and $\pi \circ \beta = x$ so that:

- $\pi \circ \beta = (\overline{\pi} \circ \sigma) \circ \beta$.

- π and $\overline{\pi} \circ \sigma$ coincide on $\beta(E)$ who generates $F(X)$ so that we must have $\pi = \overline{\pi} \circ \sigma$.

Solution 3) As mentioned in Exercise 55, $F(X)$ is the Lie subalgebra of $T_L(E)$ generated by X so that we must have $[x,y] = x \otimes y - y \otimes x$ in $F(X)$ which shows that, according to Exercise 55, $U(\mathfrak{S}) = T(\mathfrak{S})/J = T(\mathfrak{S})$. Or, we note also that this result is a consequence of the PBW theorem (see again end of Exercise 55) since the basis of $U(\mathfrak{S})$ will be in that specific case the basis of $T(\mathfrak{S})$.

58/828 ¶

Subject: Sum of objects of a category

Let $\{X_\alpha\}, \alpha \in A$, a family of objects in a category $\mathfrak{R}$. We consider the category $\tilde{\mathfrak{R}}$ whose objects are families of morphisms $\varphi_\alpha \in Mor(X_\alpha, Y), \alpha \in A$. (Y is an object of $\mathfrak{R}$ that is not the same for all objects of $\tilde{\mathfrak{R}}$). The morphisms of $\tilde{\mathfrak{R}}$ are the family of commutative diagrams of the shape:

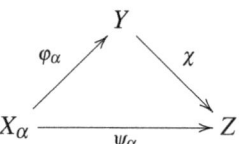

If the category $\tilde{\mathfrak{R}}$ have a universal object, then the corresponding object of $\mathfrak{R}$ is named *the sum of objects* $\bigsqcup_{\alpha \in A} X_\alpha$. The morphisms $i_\alpha : X_\alpha \to \bigsqcup_{\alpha \in A} X_\alpha$ are the *canonical injections* of the terms in the sum.

Show that the sum of a family of objects is defined in the category of sets and in the category of vector spaces on a given field.

SOLUTION:

a) Category of sets. First we need to assume that the morphisms in the category of sets will be defined as applications from a set to an other, e.g. $f \in Mor(A, B) \Leftrightarrow f : A \to B$. Now, from the family of sets $\{X_\alpha\}_{\alpha \in A}$ we can create a second family $\{X'_\alpha\}_{\alpha \in A}$ such that:

1) X'_α is a copy of X_α (e.g. there exist a bijective mapping $f_\alpha : X_\alpha \to X'_\alpha$).
2) $\forall \alpha, \beta \in A^2$, $\alpha \neq \beta \Rightarrow X'_\alpha \cap X'_\beta = \emptyset$.

We see then that $\bigcup_{\alpha \in A} X'_{\alpha}$ is partitioned by the X'_{α}. We note this set $\bigvee_{\alpha \in A} X_{\alpha}$ (*disjunctive union* of the sets X_{α}).

From here, for any object of $\mathfrak{R}$, $\{\psi_{\alpha}\}_{\alpha \in A}$, $\psi_{\alpha} : X_{\alpha} \to Z$, we can define χ by $\chi|_{X'_{\alpha}} = \psi_{\alpha} \circ f_{\alpha}^{-1}$. χ is unique and well-defined also we can see that $(\bigvee_{\alpha \in A} X_{\alpha}, \chi)$ is our universal object.

b) Category of vector spaces. The morphisms in this category are the linear mappings from one vector space to an other. Let $\{E_{\alpha}\}_{\alpha \in A}$ be a family of vector spaces. We form the direct sum $\bigoplus_{\alpha \in A} E_{\alpha}$ and we may define a morphism from Y to Z, χ, by $\chi|_{E_{\alpha}} = \phi_{\alpha}$. χ is unique and well-defined and it is straightforward to see that $(\bigoplus_{\alpha \in A} E_{\alpha}, \chi)$ is our universal object.

In a) and b), in order to ensure the existence of the objects $\bigvee_{\alpha \in A} X_{\alpha}$ and $\bigoplus_{\alpha \in A} E_{\alpha}$ we need to use the axiom of choice that claims that $\prod_{\alpha \in A} A_{\alpha} \neq \emptyset$.

In a), $\bigvee_{\alpha \in A} X_{\alpha}$ is the subset of $\prod_{\alpha \in A} A_{\alpha}$ made with the elements (or "vectors") $\{(\ldots, 0, \ldots, 0, x_{\alpha}, 0, \ldots, 0, \ldots), x_{\alpha} \in X_{\alpha}\}$ this set is also equivalent to $\{(\delta_{\alpha\beta} x_{\alpha})_{\beta \in A}, x_{\alpha} \in X_{\alpha}\}$, with the convention $\delta_{\alpha\beta} x_{\alpha} = 0$ if $\alpha \neq \beta$ and $= x_{\alpha}$ otherwise.

φ_{α} is the mapping $X_{\alpha} \to (\ldots, 0, X_{\alpha}, 0, \ldots)$.

In b), $\bigoplus_{\alpha \in A} E_{\alpha} = \{\sum_{\alpha \in A} (\delta_{\alpha\beta})_{\beta \in A} x_{\alpha}), x_{\alpha} \in E_{\alpha}\} \subset \prod_{\alpha \in A} E_{\alpha}$.

φ_{α} is the mapping $E_{\alpha} \to \ldots \oplus \vec{0} \oplus X_{\alpha} \oplus \vec{0} \oplus \ldots$

59/828

Subject: Product of objects in a category

We define the notion of a *product* of objects $\{X_\alpha\}, \alpha \in A$ in the category $\mathfrak{R}$ from the definition of a sum of objects (see exercise 58) by inverting the direction of the arrows.

In a more precise manner, we define the product $\prod_{\alpha \in A} X_\alpha$ as the sum of objects in the dual category $\tilde{\mathfrak{R}}^0$. The morphisms $p_\alpha : \prod_{\alpha \in A} X_\alpha \to X_\alpha$ are named *canonical projections* from the product into the factors.

Show that the product of objects is defined on the category of sets and on the category of vector spaces on a given field.

SOLUTION:

We have the following situation, the other objects having the same meaning as in Exercise 58:

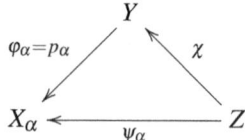

In these two special contexts, the notions of Cartesian product and the notion of product of objects are similar.

a) Category of sets.

We let $Y = \prod_{\alpha \in A} X_\alpha$.
We define χ by $(\chi(Z))_\alpha = \phi_\alpha(z), \forall \alpha \in A$. So that:

$$\chi(z) = \prod_{\alpha \in A} \phi_\alpha(z), z \in Z,$$

and the projections p_α are then defined by $p_\alpha : (\ldots, x_\alpha, \ldots) \to x_\alpha$.

b) Category of vector spaces.

We note E_α for X_α and we let $Y = \prod_{\alpha \in A} E_\alpha$.
The process is exactly the same as in a): $\prod_{\alpha \in A} E_\alpha$ is a vector space and similarly $\chi(z) = \prod_{\alpha \in A} \phi_\alpha(z), z \in E$. χ is a linear mapping $E \to \prod_{\alpha \in A} E_\alpha$.
So that in both cases the Cartesian products of the objects are the universal objects and they coincide with the notion of product of objects.

60/828

Subject: Isomorphism between the sum and the product in the category of vector spaces

Show that, in the category of vector spaces over a given field $\mathbb{K}$, the sum $\bigsqcup_{k=1}^{k=n} L_k$ and the product $\prod_{k=1}^{k=n} L_k$ of a finite number of objects are isomorphic.

SOLUTION:

From Exercises 58) and 59) this means that we must show that $\bigoplus_{\alpha \in A} E_\alpha$ and $\prod_{\alpha \in A} E_\alpha$ are isomorphic, for a family $\{E_\alpha\}_{\alpha \in A}$ of vector spaces. The demonstration of this result is straightforward since:

$$\bigoplus_{\alpha \in A} E_\alpha \approx (\dots, E_\alpha, \dots)_{\alpha \in A} = \prod_{\alpha \in A} E_\alpha.$$

61/828 $\mathscr{B}$

Subject: Tensor product

Let L_1 and L_2 two vector spaces over a field $\mathbb{K}$. We consider the category whose objects are defined by the bilinear mappings $\varphi : L_1 \times L_2 \to L$, where L is a vector space. (L is not the same for each object φ). We call morphism from an object $\varphi : L_1 \times L_2 \to L$ into an object $\psi : L_1 \times L_2 \to M$, a linear mapping $\chi : L \to M$ such that the diagram:

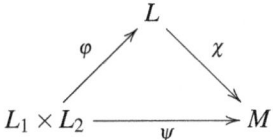

is commutative.

Show that this category has a universal repelling object:

$$\pi : L_1 \times L_2 \to L_1 \otimes L_2.$$

The vector space $L_1 \otimes L_2$ is named the Tensor product of the spaces L_1 and L_2 over the field K.

SOLUTION:

As indicated in the hint, we will consider the space $L_1 \boxtimes L_2$ made with linear combinations of symbolic products $a \boxtimes b, a \in L_1$, $b \in L_2$. $L_1 \boxtimes L_2$ is *not* equal to $L_1 \times L_2$, it is a larger set and is in fact the vector space generated by elements of $L_1 \times L_2$.

We let:

$$R(\lambda_1,\lambda_2,a_1,a_2,b) = (\lambda_1 a_1 + \lambda_2 a_2)\boxtimes b - \lambda_1(a_1 \boxtimes b) - \lambda_2(a_2 \boxtimes b)$$

and:

$$S(\mu_1,\mu_2,a,b_1,b_2) = a\boxtimes(\mu_1 b_1 + \mu_2 b_2) - \mu_1(a\boxtimes b_1) - \mu_2(a\boxtimes b_2).$$

Next we consider the vector space, $L_1 \odot L_2$ spanned by elements:

$$R(\lambda_1,\lambda_2,a_1,a_2,b)$$

and:

$$S(\mu_1,\mu_2,a,b_1,b_2)$$

[109] (taking all possible values) and then we build the space $L_1 \otimes L_2$ as the quotient space $L_1 \boxtimes L_2/L_1 \odot L_2$ [110]. We then want to prove that we have the following commutative diagram:

$$L_1 \otimes L_2 = L_1 \boxtimes L_2/L_1 \odot L_2$$

$$\pi \qquad \chi?$$

$$L_1 \times L_2 \xrightarrow{\phi} M$$

We can define $\pi : L_1 \times L_2 \to L_1 \otimes L_2$ by making $\pi(a,b) = a\otimes b$ [111], that is to say the class of $a\boxtimes b$ in $L_1 \boxtimes L_2/L_1 \odot L_2$. Following our notes in Exercise 55, the following relations hold:
$\overline{R(\lambda_1,\lambda_2,a_1,a_2,b)} = 0$ and $\overline{S(\mu_1,\mu_2,a,b_1,b_2)} = 0$ will hold in $L_1 \otimes L_2$.
So that:

---

[109] The set of elements $R(\lambda_1,\lambda_2,a_1,a_2,b)$ and $S(\mu_1,\mu_2,a,b_1,b_2)$ is not a vector space
[110] $L_1 \otimes L_2$ is a vector space because it is the quotient of two vector spaces
[111] The image of π is not $L_1 \times L_2$, e.g. if $x \in L_1 \times L_2$ then $\pi(x)$ is not necessarily equal to an element $a\otimes b$, $a \in L_1$, $b \in L_2$. In fact we have that $\pi(x) = \sum_{i=1}^{N} a_i \otimes b_i$

$$(\lambda_1 a_1 + \lambda_2 a_2) \boxtimes b = \overline{\lambda_1 a_1 \boxtimes b} + \overline{\lambda_2 a_2 \boxtimes b}$$

and similarly:

$$\overline{a \boxtimes (\mu_1 b_1 + \mu_2 b_2)} = \overline{\mu_1 a \boxtimes b_1} + \overline{\mu_2 a \boxtimes b_2}.$$

As $\overline{a \boxtimes b} = a \otimes b$, we have that π is bilinear, e.g. that $(\lambda_1 a_1 + \lambda_2 a_2) \otimes b - \lambda_1 (a_1 \otimes b) - \lambda_2 (a_2 \otimes b) = 0$ and that $a \otimes (\mu_1 b_1 + \mu_2 b_2) - \mu_1 (a \otimes b_1) - \mu_2 (a \otimes b_2) = 0$.

In order to factorize ϕ by π we would like to show that ker π is included in ker ϕ but that will not work directly here. Instead we must introduce the space $L_1 \boxtimes L_2$ as follows.

We define the application $\varphi : L_1 \times L_2 \to L_1 \boxtimes L_2$ by $\varphi(a,b) = a \boxtimes b$. Then we want to see if we can factor π by φ and also factor ϕ by φ like in the following diagram:

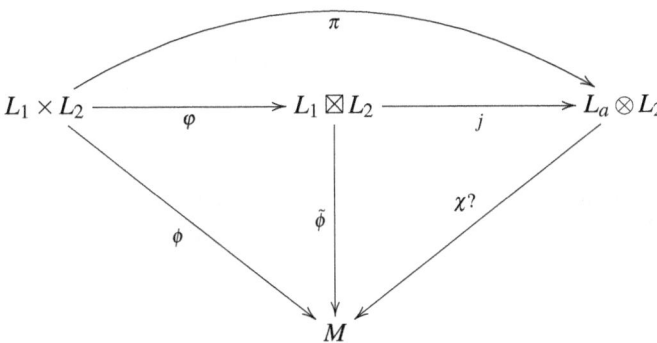

We can try to define $\tilde{\phi}$ over $L_1 \boxtimes L_2$ by:

$$\tilde{\phi}(\sum_i \lambda_i a_i \boxtimes b_i) = \sum_i \lambda_i \phi(a_i, b_i).$$

Then we see that $\tilde{\phi} \circ \varphi(a,b) = \tilde{\phi}(a \boxtimes b) = \phi(a,b)$ so that $\phi = \tilde{\phi} \circ \varphi$. It is immediate to check that $\tilde{\phi}$ is an homomorphism.

In the same way π is factored by φ through: $\pi = j \circ \varphi$, where j is the canonical mapping from $L_1 \boxtimes L_2 \to L_1 \boxtimes L_2 / L_1 \odot L_2 = L_1 \otimes L_2$.

We now can note that $\tilde{\phi}(L_1 \odot L_2) = 0$, indeed, if $x = (\lambda_1 a_1 + \lambda_2 a_2) \boxtimes b - \lambda_1(a_1 \boxtimes b) - \lambda_2(a_2 \boxtimes b)$, then:

$$
\begin{aligned}
\tilde{\phi}(x) &= \tilde{\phi}((\lambda_1 a_1 + \lambda_2 a_2) \boxtimes b - \lambda_1(a_1 \boxtimes b) - \lambda_2(a_2 \boxtimes b)), \\
&= \phi(\lambda_1 a_1 + \lambda_2 a_2, b) - \lambda_1 \phi(a_1, b) - \lambda_2 \phi(a_2, b), \\
&= 0,
\end{aligned}
$$

since ϕ is bilinear.

The same occurs for $y = a \boxtimes (\mu_1 b_1 + \mu_2 b_2) - \mu_1(a \boxtimes b_1) - \mu_2(a \boxtimes b_2)$, e.g. $\tilde{\phi}(y) = 0$.

So that $\ker j \subset \ker \tilde{\phi}$, hence $\tilde{\phi}$ can be factored by j in, let us say, $\tilde{\phi} = \chi \circ j$.

Then our diagram will be commutative:

$$
\begin{aligned}
\chi \circ \pi &= (\chi \circ j) \circ \varphi, \\
&= \tilde{\phi} \circ \varphi, \\
&= \phi.
\end{aligned}
$$

We have finally proved that the tensor product of L_1 by L_2 over the field $\mathbb{K}$ is well defined as the universal object of our category.

62/828 ¶

Subject: Torsion product

Let G_1 and G_2 be finite Abelian groups. We consider the category of all applications:

$$\varphi : G_1 \times G_2 \to G.$$

Where G is an Abelian group (which is not the same for each objects) and who are homomorphism for each variables. The morphisms will be the commutative diagrams of the shape:

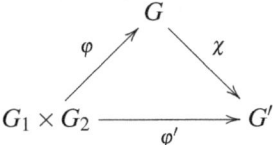

Where χ is an homomorphism and G' is a group. Show that the category have a universal object:

$$G_1 \times G_2 \to Tor(G_1, G_2)$$

($Tor(G_1, G_2)$ is called the *Torsion product* of the two groups). Calculate $Tor(C_m, C_n)$ where C_m is the cyclic group of order m.

SOLUTION:

Let us first consider the case where $G_1 = \mathbb{Z}/n\mathbb{Z}$ and $G_2 = \mathbb{Z}/m\mathbb{Z}$. $(m,n) \in \mathbb{N}^*$.

We consider the mapping φ: $\mathbb{Z}/n\mathbb{Z} \times \mathbb{Z}/m\mathbb{Z} \to \mathbb{Z}/d\mathbb{Z}$ defined by: $\varphi(a,b) = ab$ and where $d = gcd(m,n)$.

This mapping is a bi-homomorphism because $x \to ax$ is an homorphism from $\mathbb{Z}/m\mathbb{Z}$ in $\mathbb{Z}/d\mathbb{Z}$ and $y \to yb$ is an homomorphism from $\mathbb{Z}/n\mathbb{Z}$ in $\mathbb{Z}/d\mathbb{Z}$. φ is surjective because $\varphi(\mathbb{Z}/n\mathbb{Z},1) \cup \varphi(1,\mathbb{Z}/m\mathbb{Z}) = \mathbb{Z}/d\mathbb{Z}$ $(d \leq \max(m,n))$.

If G' is an abelian finite group and φ' a bi-homomorphism from $\mathbb{Z}/n\mathbb{Z} \times \mathbb{Z}/m\mathbb{Z} \to G'$, then, following the hint, we try to prove that $\mathbb{Z}/d\mathbb{Z}$ is our universal object $\mathrm{Tor}(\mathbb{Z}/n\mathbb{Z},\mathbb{Z}/n\mathbb{Z})$.

For this, we must prove that there exist a unique group-homomorphism $\chi : \mathbb{Z}/d\mathbb{Z} \to G'$ such that $\chi \circ \varphi = \varphi'$.

We first look what happens when, for example, $n = 21$, $m = 14$. Then $d = gcd(21,14) = 7$, and if we take two sample points, for example, $(a,b) = (6,3)$ and $(c,d) = (5,2)$ in $\mathbb{Z}/21\mathbb{Z} \times \mathbb{Z}/14\mathbb{Z}$, then we have $\varphi(a,b) = \varphi(6,3) = 18(7) = 4$ and $\varphi(c,d) = \varphi(5,2) = 10(7) = 3$, then we must find $\chi \in Hom(\mathbb{Z}/7\mathbb{Z},G')$ such that $\chi \circ \varphi(a,b) = \varphi'(a,b)$ and $(\chi \circ \varphi)(c,d) = \varphi'(c,d)$.

That is to say: $\chi(4) = \varphi'(6,3)$ and $\chi(3) = \varphi'(5,2)$

From this, it looks possible to define uniquely χ because we will have $\chi(a) = \varphi'(1,a) = \varphi(a,1) \forall a \in \mathbb{Z}/7\mathbb{Z}$ so that in fact χ (if it exist) is uniquely defined by:

$\chi(1) = \varphi'(1,1),$
$\chi(2) = \varphi'(2,1),$
...
$\chi(7) = \varphi'(7,1).$

The only problem is therefore the existence of χ. From its definition, χ — if it exists — will be an homomorphism because $\chi(a) = \varphi'(a,1)$ and $\chi(b) = \varphi'(b,1)$ will ensure that $\chi(a) + \chi(b) = \varphi'(a,1) + \varphi'(b,1) = \varphi'(a+b,1) = \chi(a+b)$.

Therefore, we will have unicity of the definition if and only if $ab = cd$ in $\mathbb{Z}/7\mathbb{Z}$ always imply that $\varphi'(a,b) = \varphi'(c,d)$.

For example, we have $7 \times 2 = 7 \times 3 = 0$ in $\mathbb{Z}/7\mathbb{Z}$ so that we must check that $\varphi'(7,2) = \varphi'(7,3)$.

Additional properties. We have that $\varphi'(21,a) = \varphi'(7 \times 3, a) = 3 \times \varphi'(7,a) \forall a \in \mathbb{Z}/14\mathbb{Z}$ ($x \to \varphi'(x,a)$ being an homorphism $\mathbb{Z}/21\mathbb{Z} \to G'$, $\forall a \in \mathbb{Z}/14\mathbb{Z}$) so that $\varphi'(7,a) = 0, \forall a \in \mathbb{Z}/14\mathbb{Z}$ and we have then $\varphi'(7,2) = \varphi'(7,3)$. So that the fact that φ' is a bi-homomorphism seems to ensure that $ab = cd \mod (\mathbb{Z}/7\mathbb{Z}) \Rightarrow \varphi'(a,b) = \varphi'(c,d)$.

In the general case, we define χ by:

$$\chi(a) = \varphi'(a,1), \forall a \in \mathbb{Z}/d\mathbb{Z}$$

and we need to check the non-contradiction of our definition by making sure that: $\forall (a,b)$ and $\forall (c,d) \in \mathbb{Z}/n\mathbb{Z} \times \mathbb{Z}/m\mathbb{Z}$, $ab = cd (\mod \mathbb{Z}/d\mathbb{Z}) \Rightarrow \varphi'(a,b) = \varphi'(c,d) \mod \mathbb{Z}/d\mathbb{Z}$. As we have seen in the example, we can use the fact that φ' is a bi-homomorphism so that:
$\varphi'(x, ab - cd) = \varphi'(x, k \times d) = k \times \varphi'(x,d) = 0$, so that $\forall x \in \mathbb{Z}/n\mathbb{Z}$, $\varphi'(x, ab) - \varphi'(x.cd) = 0$ or equivalently $a\varphi'(x,b) - c\varphi'(x,d) = 0$ or $\varphi'(ax, b) - \varphi'(cx, d) = 0$.

If we make $x = 1$ then we have: $\varphi'(a,b) = \varphi'(c,d)$.

We have that $\chi(ab) = \varphi'(a,b), \forall (a,b) \in \mathbb{Z}/n\mathbb{Z} \times \mathbb{Z}/m\mathbb{Z}$ and $\chi(ab) = \varphi'(ab,1) = \varphi'(1,ab)$ so that:
$\forall (a,b)$ and $\forall (c,d) \in \mathbb{Z}/n\mathbb{Z} \times \mathbb{Z}/m\mathbb{Z}$, $\chi(ab + cd) = \varphi'(ab + cd, 1) = \varphi'(ab, 1) + \varphi'(cd, 1) = \chi(ab) + \chi(cd)$.

We even have more than that, since $\chi(ab + cd) = a\chi(b) + c\chi(d) \forall (a,b), \forall (c,d) \in \mathbb{Z}/n\mathbb{Z} \times \mathbb{Z}/m\mathbb{Z}$.

So that $Tor(\mathbb{Z}/n\mathbb{Z}, \mathbb{Z}/m\mathbb{Z}) = \mathbb{Z}/d\mathbb{Z}$, $d = gcd(n,m)$.

In the general case, we will use the fact, as indicated in the hint, that all Abelian finite groups are, up to one group-isomorphism, direct sums of finite cyclic groups, e,g that for any finite Abelian group G, we can find $m_1, \ldots, m_N$ such that:

$$G = \bigoplus_{i=1}^{i=N} \mathbb{Z}/m_i\mathbb{Z}.$$

We also note, as suggested in the hint, that we can define a functor Tor_{G_1} from the category $\mathfrak{C}$ of Abelian groups to the category $\mathfrak{C}$ of Abelian group by $Tor_{G_1} : ob(\mathfrak{C}) \to ob(\mathfrak{C})$, $G \to Tor_{G_1}(G) = Tor(G_1, G)$, so that we may define the action of the functor Tor_{G_1} on $\mathrm{Mor}(\mathfrak{C})$ by: if $f_{uv} \in \mathrm{Mor}(U,V)$, $(U,V) \in \mathrm{ob}(\mathfrak{C})^2$, then we define canonically an homomorphism $f_{uv} : G_1 \times U \to G_1 \times V$ and we define $Tor_{G_1}(f_{uv}) = \chi_{uv}$, e.g. as the only group-homomorphism (it exists, following the universal property of the Torsion group) such that the following diagram be commutative:

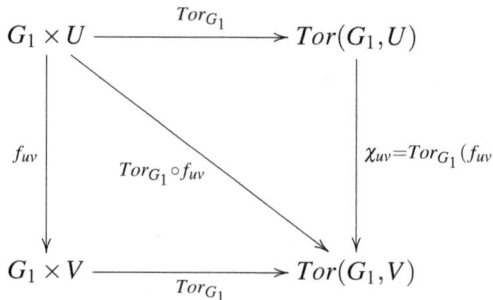

Let us look at the action of Tor_{G_1} on 3 objects (U,V,W) of $\mathfrak{C}$ (canonically identified as $G_1 \times U$, $G_1 \times V$, $G_1 \times W$). Then both $\chi_{vw} \circ \chi_{uv}$ and χ_{uw} are group-homomorphisms that make the following diagram commutative:

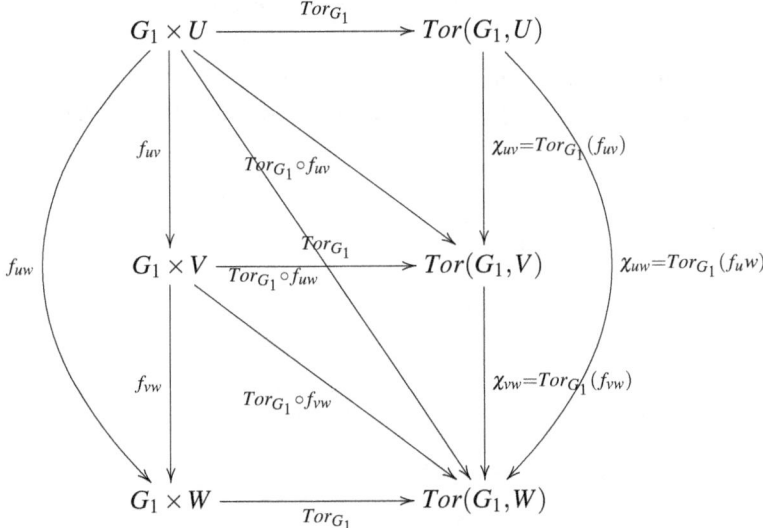

We therefore must have $\chi_{uw} = \chi_{vw} \circ \chi_{uv}$ by definition of Tor as a universal object. Finally, Tor is a functor from the category $\mathfrak{C}$ to the category $\mathfrak{C}$.

We can also define Tor_{G_2} as a functor from $\mathfrak{C}$ to $\mathfrak{C}$ by making $U \rightarrow Tor(G_2, U)$. So that we may view the correspondence from $\mathfrak{C}^2$ to $\mathfrak{C}$: $(U, V) \rightarrow Tor(U, V)$ as a bifunctor .

We wish to prove next that Tor is an additive bifunctor (such like The tensor product Functor for example), e.g. that the correspondence $Mor(\mathfrak{C}) \times Mor(\mathfrak{C}) \rightarrow MorTor(\mathfrak{C}, \mathfrak{C}) = Mor(\mathfrak{C})$ is bi-additive.

That is to say that we must prove that: $\forall (\varphi, \varphi') \in Mor(\mathfrak{C})$ and $\forall (\sigma, \sigma') \in Mor(\mathfrak{C})$ then:

$$Tor(\varphi + \varphi', \sigma) = Tor(\varphi, \sigma) + Tor(\varphi', \sigma)$$

and

$$Tor(\varphi, \sigma + \sigma') = Tor(\varphi, \sigma) + Tor(\varphi, \sigma').$$

In fact this is rather straightforward because if f_{uv} and f'_{uv} are two morphism from U to V then $\chi_{uv} + \chi'_{uv}$ will make the following diagram commutative:

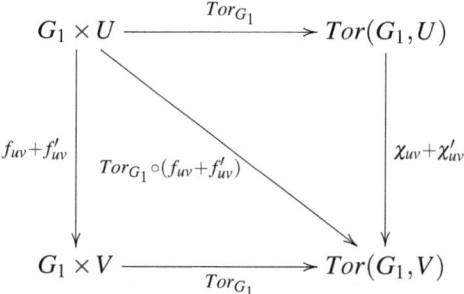

Then by unicity of the universal object, we must have:

$$Tor_{G_1}(f_{uv} + f'_{uv}) = Tor_{G_1}(f_{uv}) + Tor_{G_1}(f'_{uv}).$$

The same occurs for Tor_{G_2} so that Tor is an additive bifunctor.
Then we must have $Tor_{G_1}(U \oplus V) = Tor_{G_1}(U) \oplus Tor_{G_1}(V)$.
Indeed the morphisms of $U \oplus V$ are $Mor(U) \oplus Mor(V)$ and then:

$$Tor_{G_1}(Mor(U \oplus V)) = MorTor_{G_1}(U) \oplus MorTor_{G_1}(V)$$

and then the group $Tor_{G_1}(U) \oplus Tor_{G_1}(V)$ is a solution to the universal object as shown in the following diagram:

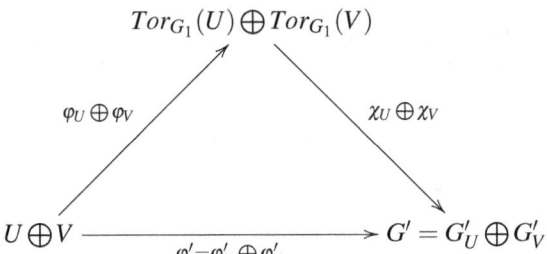

So this means that $Tor(U \oplus V) = Tor(U) \oplus Tor(V) \ \forall (U,V) \in ob(\mathfrak{C})^2$.

So if we use the decomposition of any two finite Abelian group G1 and G 2 in direct sum of cyclic groups[112], e.g. let us say:

$$G_1 = \bigoplus_{i=1}^{i=N_1} \mathbb{Z}/n_i\mathbb{Z} \text{ and } G_2 = \bigoplus_{i=1}^{i=N_2} \mathbb{Z}/m_i\mathbb{Z}.$$

We will have:

$$Tor(G_1, G_2) = \bigoplus_{i=1,...N_1; j=1,...N_2} \mathbb{Z}/d_{ij}\mathbb{Z}, \ d_{ij} = gcd(n_i, m_j).$$

.

Notes

[i] The torsion group of a group A, named $Tor(A)$ is defined as the set of elements of A which have a finite period, that is to say the set of $x \in A$ such as $nx = 0$ for some $n \in \mathbb{N}$ (A being noted additively).

The torsion group and the torsion product $A \star B$ between two Abelian groups are intimately connected since we can show that the torsion product of A and B only depends of the torsion groups of A and B, e.g.

$$A \star B = Tor(A) \star Tor(B).$$

[112]This is the application to finite Abelian groups of the fundamental theorem for finitely generated Abelian groups, see [?]

[ii] The Torsion product and the Tor functor (and bifunctor) associated to it can be defined in a much more general context[113]:

First, we can define, in the context of Abelian groups, the notion of a *free resolution* of A. This is a short exact sequence containing two free groups F_1 and F_2 and two group-homomorphisms f_1 and f_2 as follows:

$$0 \longrightarrow F_1 \xrightarrow{f_1} F_2 \xrightarrow{f_2} A \longrightarrow 0 .$$

By an exact sequence, we mean that we have the following properties:

$$\begin{cases} \ker(f_1) = 0, \\ Im(f_1) = \ker(f_2), \\ Im(f_2) = A. \end{cases}$$

We can define also the tensor product $\otimes$ on a more general context than in Exercise 61, that is to say the tensor product $\otimes_R$ between modules over a ring R. Abelian groups are all $\mathbb{Z}$-modules so, if we process the free resolution sequence through the functor $(.) \rightarrow B \otimes_{\mathbb{Z}} (.)$, we get a new sequence:

$$0 \longrightarrow B \otimes F_1 \xrightarrow{id \otimes f_1} B \otimes F_2 \xrightarrow{id \otimes f_2} B \otimes A \longrightarrow 0 .$$

We define $Tor(A,B)$ as the 1-th homology of this derived sequence , that is to say, here, as the group $\ker(id \otimes f_1)$:

$$A \star B = \ker(id \otimes f_1).$$

We can show that this definition does not depend of the choice of the free resolution of A. For example if we want to compute $L \star \mathbb{Z}/n\mathbb{Z}$, we can use the following sequence:

$$0 \longrightarrow \mathbb{Z} \xrightarrow{f_n} \mathbb{Z} \xrightarrow{j} \mathbb{Z}/n\mathbb{Z} \longrightarrow 0$$

where f_n and j are defined by:

$$\mathbb{Z} \rightarrow \mathbb{Z},$$
$$x \rightarrow f_n(x) = nx.$$

$$\mathbb{Z} \rightarrow \mathbb{Z}/p\mathbb{Z},$$
$$x \rightarrow j(x) = x(mod n).$$

We check that this is and exact sequence:

---

[113] We can show that this definition is equivalent with the definition given in the exercise

$$\begin{cases} \ker(f_n) = 0, \\ Im(f_n) = n\mathbb{Z} = \ker(j), \\ Im(j) = \mathbb{Z}/n\mathbb{Z}. \end{cases}$$

Then we consider the derived sequence obtained by the action of the Tensor functor $(.) \to L \otimes_{\mathbb{Z}} (.)$ and we get:

$$0 \longrightarrow L \otimes_{\mathbb{Z}} \mathbb{Z} \xrightarrow{id \otimes_{\mathbb{Z}} f_n} L \otimes_{\mathbb{Z}} \mathbb{Z} \xrightarrow{j \otimes_{\mathbb{Z}} f_n} L \otimes_{\mathbb{Z}} \mathbb{Z}/n\mathbb{Z} \longrightarrow 0 \ .$$

Since $L \otimes_{\mathbb{Z}} \mathbb{Z} = L$ (L is a $\mathbb{Z}$-module), our derived sequence reduce itself to:

$$0 \longrightarrow L \xrightarrow{f_n'} L \xrightarrow{j'} L \otimes_{\mathbb{Z}} \mathbb{Z}/n\mathbb{Z} \longrightarrow 0 \ ,$$

f_n' being defined by:

$$L \to L,$$
$$x \to f_n'(x) = nx.$$

Then we can easily see now that $Tor(L, \mathbb{Z}/n\mathbb{Z}) = L \star \mathbb{Z}/n\mathbb{Z} = \ker(f_n') = \{y \in L, ny = 0\}$

[iii] The Tor functor can be defined in, even a much more general context: Tor(A,B) may be defined if A and B are R-modules, as the 1th homology of the derived sequence obtained by applying the tensor functor $(.) \to B \otimes (.)$ to a free (or projective) resolution of A. We also may define the jth Torsion product as the jth homology of this derived sequence. The Tor bifunctor is then being defined as the *derived functor* of the Tensor bifunctor.

E.g. we consider a free resolution of A:

$$0 \longrightarrow E_1 \xrightarrow{f_1} E_2 \xrightarrow{f_2} \dots \xrightarrow{f_{n-1}} E_n \xrightarrow{f_n} A \longrightarrow 0$$

then, by applying the tensor product with B on the left, we get:

$$0 \longrightarrow B \otimes E_1 \xrightarrow{id \otimes f_1} B \otimes E_2 \xrightarrow{id \otimes f_2} \dots \xrightarrow{id \otimes f_{n-1}} B \otimes E_n \xrightarrow{id \otimes f_n} B \otimes A \longrightarrow 0 \ .$$

We define $Tor_j(A, B)$ by the following formula:

$$Tor_j(A, B) = \ker(id \otimes f_{j+1})/Im(id \otimes f_j).$$

This does not depend of the choice of the free resolution.

63/828 ¶

Subject: Direct and Inverse limits

Let A be a filtering set, $\mathfrak{R}$ a category. For every $\alpha \in A$, we consider $X_\alpha \in Ob(\mathfrak{R})$ and for every couple $\alpha \leq \beta$, we consider a morphism $\varphi_{\alpha,\beta} \in Mor(X_\alpha, X_\beta)$. The following diagram:

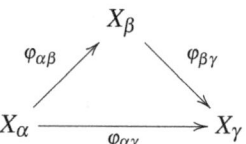

being commutative for every triple $\alpha \leq \beta \leq \gamma$.

We consider the category $\mathfrak{R}_A$ whose objects are the families of morphisms $\{\varphi_\alpha : X_\alpha \to X\}_{\alpha \in A}$ compatible with $\varphi_{\alpha,\beta}$ where X is an object of $\mathfrak{R}$(that is not the same for each family). We name morphism from $\{\varphi_\alpha : X_\alpha \to X\}_{\alpha \in A}$ into $\{\psi_\alpha : X_\alpha \to Y\}_{\alpha \in A}$ a morphism $\xi \in Mor(X,Y)$ such that, for every $\alpha \in A$, the diagram:

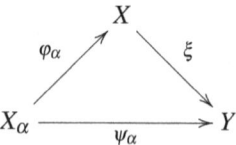

is commutative.

The universal object (if any) of the category $\mathfrak{R}_A$ is named the *inductive limit* of the family $\{X_\alpha\}_{\alpha \in A}$. The dual notion of *projective limit* is defined as the universal object in $\mathfrak{R}_A^0$.

Show that:

a) The additive group of the rationals is the inductive limit (or direct limit) of an enumerable family of a groups of entire numbers.

b) $\mathbb{Z}_p$, the ring of p-adics (see Exercise 42) is the projective linit (or inverse limit) of the residual rings $\mathbb{Z}/p^n\mathbb{Z}$.

SOLUTION:

a) Let us show that $(\mathbb{Q}, +)$ is the universal object in our category. We take the base category as the category $\mathfrak{C}$ of commutative additive groups.

Then we consider $\mathbb{Z}_\alpha$, a copy of $\mathbb{Z}$ indexed by $\alpha \in \mathbb{N}$ and $\beta \geq \alpha \Leftrightarrow \alpha | \beta$. We define the family $\{f_{\alpha\beta}\}$ by $f_{\alpha\beta}(k) = \frac{\beta}{\alpha}k$, $\forall k \in \mathbb{Z}$. We then have a category $\mathfrak{C}_\mathbb{N}$ whose objects are the directed families (f_u, A), A being a commutative additive group and the family $\{f_u\}$ being compatible with the family $\{f_{\alpha\beta}\}$.

We define the family $f_\alpha : \mathbb{Z}_\alpha \to (\mathbb{Q}, +)$ by $f_\alpha(k) = \frac{k}{\alpha}$. We check immediately that $f_\alpha = f_\beta \circ f_{\alpha\beta}$: indeed, $f_\beta \circ f_{\alpha\beta}(k) = f_\beta(\frac{\beta}{\alpha}k) = \frac{k}{\alpha} = f_\alpha(k)$.

We then have the following situation where (g_α, B) is an object of $\mathfrak{C}_\mathbb{N}$ and $(f_\alpha, (\mathbb{Q}, +))$ the object of $\mathfrak{C}_\mathbb{N}$ that we need to check is the universal object :

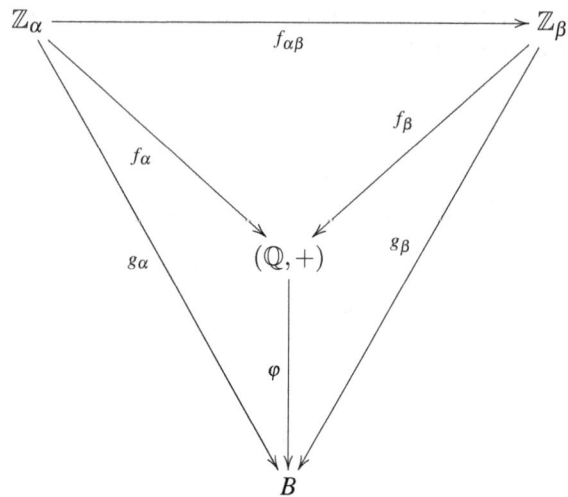

The family $\{g_\alpha\}$ is such that:

$$g_\beta \circ f_{\alpha\beta} = g_\alpha \ \forall(\alpha,\beta).$$

We need to find a group-homomorphism φ, from $(\mathbb{Q},+)$ to $(B,+)$ such that the above diagram is commutative.

We get the following formula that link g_α and g_β:

$$g_\alpha(k) = g_\beta(\frac{\beta}{\alpha}k) \ \forall(\alpha,\beta,k), \alpha|\beta.$$

We know that g_α is a group-homomorphism from $\mathbb{Z}$ into a group B. One must have $g_\alpha(n.1) = n.g_\alpha(1)$ $(\forall n \in \mathbb{N})$ (this is trivially verified by recurrence). Since $g_\alpha(-n.1) = -g_\alpha(n.1) = -n.g_\alpha(1)$, we can write: $g_\alpha(k) = k.g_\alpha(1)$.

Thus the morphisms $g_\alpha : \mathbb{Z}_\alpha \to B$ are uniquely determined by $g_\alpha(1)$. If φ exists, it is defined by $\varphi \circ f_\alpha = g_\alpha$:

$$\varphi \circ f_\alpha(k) = g_\alpha(k), \ \forall \alpha \in \mathbb{N}, \forall k \in \mathbb{Z}_\alpha.$$

That is to say equivalently:

$$\varphi(\frac{k}{\alpha}) = k.g_\alpha(1), \quad \forall \alpha \in \mathbb{N}, \forall k \in \mathbb{Z}_\alpha.$$

This relation ensures us that if φ exists it is uniquely defined.

What we need to check, in order to be sure of the existence of φ, is that $\frac{k}{\alpha} = \frac{k'}{\alpha'} \Rightarrow k'g_{\alpha'}(1) = kg_\alpha(1)$.

We have that $g_\alpha(1) = \frac{\beta}{\alpha}g_\beta(1)$ if $\alpha|\beta$. Here we can assume that $\Delta(k, \alpha) = 1$ so that $k|k'$ and $\alpha|\alpha'$: then we have $g_\alpha(1) = \frac{\alpha'}{\alpha}g_{\alpha'}(1)$ or $k.g_\alpha(1) = k.\frac{\alpha'}{\alpha}.g_{\alpha'}(1) = k'.g_{\alpha'}(1)$. Then φ exist and is unique.

We next need to check that φ is a group-homomorphism: $\varphi(\frac{k}{\alpha} + \frac{k'}{\alpha'}) = k.g_\alpha(1) + k'.g_{\alpha'}(1) = \alpha.g_{\alpha\alpha'}(k) + \alpha'.g_{\alpha\alpha'}(k') = (k\alpha + k'\alpha')g_{\alpha\alpha'}(1) = \varphi(\frac{k\alpha' + k'\alpha}{\alpha\alpha'}) = \varphi(\frac{k}{\alpha}) + \varphi(\frac{k'}{\alpha'})$. So that φ is an homomorphism $(\mathbb{Q}, +) \to (B, +)$. And then from its definition, $(\mathbb{Q}, +), f_\alpha$ is a solution to the universal object problem in our category. Then we have: $\varinjlim \mathbb{Z}_\alpha = (\mathbb{Q}, +)$. We note that $(\mathbb{Q}, +)$ can also be expressed as: $\varinjlim \frac{1}{\alpha}\mathbb{Z}_\alpha = (\mathbb{Q}, +)$ (with f_α and $f_{\alpha\beta}$ defined by: $f_\alpha(k) = k$ and $f_{\alpha\beta}(k) = k$).

b) We need to show that we have the following commutative diagram, where φ is uniquely defined for the object $(A, \{g_n\}_{n\in\mathbb{N}})$ and where $n > m$, A is an object from the base category (for example, from the category of groups) and φ is a morphism $A \to \mathbb{Z}_p$:

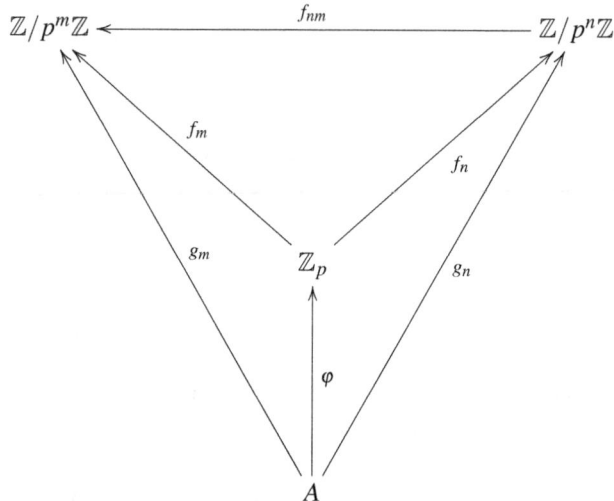

First we note that we can define f_{nm} as the canonical surjection $\mathbb{Z}/p^n\mathbb{Z} \to \mathbb{Z}/p^m\mathbb{Z}$, e.g. $f_{nm}(\tilde{x}) = x(p^m)$ [114].

Following the hint, we have a look at the injective mapping $i:\mathbb{Z} \to \mathbb{Z}_p$.

We note j_{p^n} the canonical projection $\mathbb{Z} \to \mathbb{Z}/p^n\mathbb{Z}$ defined by $j_{p^n}(x) = x(p^n)$, j'_{p^n} the canonical projection defined by: $\mathbb{Z}_p \to \mathbb{Z}_p/p^n\mathbb{Z}_p$, $j'_{p^n}(x) = x(p^n)$ and i' the injection $\mathbb{Z}/p^n\mathbb{Z} \to \mathbb{Z}_p/p^n\mathbb{Z}_p$ induced by i.

We see that for every $y \in \mathbb{Z}_p$ such that $y = \sum_{i=0}^{i=n-1} b_i p^i + \sum_{j=n}^{j=\infty} b_j p^j$, then we have $x = y(p^n)$ in $\mathbb{Z}_p$ is and only if $a_i = b_i$ for $i = 0,\ldots,p-1$, e.g. j_{p^n} "truncate" y by removing all its component of order $\geq n$: $j_{p^n}(\sum_{i=0}^{i=n-1} b_i p^i + \sum_{j=n}^{j=\infty} b_j p^j) = \sum_{i=0}^{i=n-1} a_i p^i$.

This implies that $\mathbb{Z}_p/p^n\mathbb{Z}_p$ is being described by the following p^n distinct representants from $\mathbb{Z}_p$: $\{\tilde{x}, x = \sum_{i=0}^{i=n-1} a_i p^i, 0 \leq a_i \leq p-1$. We will consider that it is clear whenever we are talking of a representant or not so that we might not use the symbol $\tilde{x}$ in what follows.

---

[114] $\tilde{x}$ being the class of $x \in N$ in $\mathbb{Z}_p/p^n\mathbb{Z}_p$

If $x > 0$. We assume that x has the following representation in base p: $x = \sum_{i=0}^{i=n-1} a_i p^i$ (in $\mathbb{Z}$). Then i is the mapping $i : x \to \sum_{i=0}^{i=n-1} a_i p^i$, $0 \leq a_i \leq p - 1$. $j'_{p^n} \circ i(x) = j'_{p^n}(\sum_{i=0}^{i=n-1} a_i p^i) = \sum_{i=0}^{i=n-1} a_i p^i = i' \circ j_p^n(x)$.

If $x < 0$ then if we assume that $-x = \sum_{i=0}^{n-1}(a_i)p^i$ is the decomposition of $-x$ in base p, i is described by the following formula: $i(x) = \sum_{i=0}^{n-1}(p - 1 - a_i)p^i + \sum_{n}^{\infty}(p - 1)p^i + 1$. Then $j'_n \circ i(x) = \sum_{i=0}^{n-1}(p - 1 - a_i)p^i + 1$ and $j'_{p^n} \circ i(0) = 0$.

We can consider $x' = p^n + x = p^n - \sum_{i=0}^{n-1} a_i p^i$ and we will have $j_{p^n}(x) = j_{p^n}(x') = p^n - \sum_{i=0}^{n-1} a_i p^i$.

But $j'_{p^n}(i(x)) = \sum_{i=0}^{n-1}(p - 1 - a_i)p^i + 1 = \sum_{i=0}^{n-1}(p - 1)p^i + 1 - \sum_{i=0}^{n-1} a_i p^i = (p - 1)\frac{1 - p^n}{1 - p} + 1 - \sum_{i=0}^{n-1} a_i p^i = p^n - \sum_{i=0}^{n-1} a_i p^i = i'(j_{p^n}(x))$.

So that in both cases, we have $i' \circ j_{p^n} = j'_{p^n} \circ i$, e.g. the following diagram is commutative:

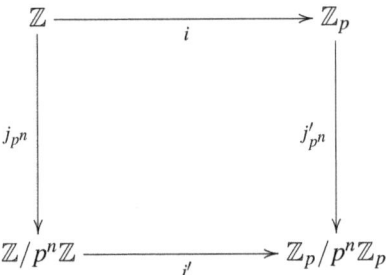

i' is an homomorphism and, since both $\mathbb{Z}/p^n\mathbb{Z}$ and $\mathbb{Z}_p/p^n\mathbb{Z}_p$ has p^n elements, i' is a bijection.

That means that $i, i : \mathbb{Z} \to \mathbb{Z}_p$ induces an isomorphism $\mathbb{Z}/p^n\mathbb{Z} \approx \mathbb{Z}_p/p^n\mathbb{Z}_p$.

This leads to defining the mappings $\{f_n\}_{n \in N}$ by the canonical mapping $\mathbb{Z}_p \to \mathbb{Z}_p/p^n\mathbb{Z}_p$ modulo i', e.g. by:

$$f_n(x) = i'^{-1} \circ j'_{p^n}.$$

And indeed, we check that $f_m = f_{nm} \circ f_n$, because if $x = \sum_{i=0}^{i=\infty} a_i p^i$, then $f_n(x) = i'^{-1}(\widetilde{\sum_{i=0}^{i=n-1} a_i p^i}$ [115]) that is to say $f_n(x) = \sum_{i=0}^{i=n-1} a_i p^i$ and $f_m(x) =$

---

[115] The class being in $\mathbb{Z}_p/p^n\mathbb{Z}_p$

$\sum_{i=0}^{i=m-1} a_j p^j$. Then $f_n m \circ f_n(x) = \sum_{i=0}^{i=n-1} a_i p^i (p^m) = \sum_{i=0}^{i=m-1} a_i p^i = f_m(x)$.

Now that we have checked that $(\mathbb{Z}_p, \{f_n\}_{n \in N})$ is an object of our category, we must check that it is universal. For this we must first try to define φ for any other object $A, \{g_n\}_{n \in N})$.

We must have $f_m \circ \varphi(a) = g_m(a), \forall a \in A$. This defines uniquely $\varphi(a)$. Indeed, if $\varphi(a) = \sum_{i=0}^{\infty} \varphi_i(a) p^i$ then we have $\sum_{i=0}^{m-1} \varphi_i(a) p^i = g_m(a)$, $\forall m$. But this is equivalent to the recursive relation in $\mathbb{Z}_p$:

$$\varphi_m(a) p^{m-1} = g_m(a).$$

Let r_m be the inverse, in $\mathbb{Z}_p$, of p^{m-1}, then φ is completely defined by:

$$\varphi_m(a) = r_m^{-1} g_m(a), \ \forall a \in A, \forall m \in \mathbb{N}.$$

The only thing we need to check now so far is that $g_m = f_{nm} \circ g_n$ $\forall (m,n) \in \mathbb{N}^2, n > m$.

But the demonstration of this result is straightforward, since:

$$
\begin{aligned}
f_m &= f_{nm} \circ f_n \Rightarrow \\
&\quad f_m \circ \varphi = f_{nm} \circ f_n \circ \varphi \Rightarrow \\
&\quad (f_m \circ \varphi) = f_{nm} \circ (f_n \circ \varphi) \Rightarrow \\
&\quad g_m = f_{nm} \circ g_n.
\end{aligned}
$$

Then $(\mathbb{Z}_p, \{f_n\}_{n \in N})$ is the universal object and we can write: $\mathbb{Z}_p = \varprojlim \mathbb{Z}/p^n \mathbb{Z}$.

Alternative solution.

From [?] we have the following result that claims that if we have a sequence $\{H_r\}$ of normal subgroups such that H_r is normal in H_{r+1} for all $r \geq 0$ (e.g. a tower), the we may define the notion of a *Cauchy sequence* $\{x_n\}_{n \in \mathbb{N}}$ in G, in a context of group theory, by the property that for any given H_r we can find N such that $(m,n) \geq N$ will imply that $x_n x_m^{-1} \in H_r$. We may define similarly a null sequence by the property that for a given H_r, we can find N such that $n \geq N$ will imply that $x_n \in H_r$. Then the set of Cauchy sequences will form a group $\mathscr{C}$ for pointwise product (e.g. $\{x_n\}_n \times \{y_n\}_n = \{x_n \times y_n\}_n$) and the set of null sequence will form a normal subgroup $\mathscr{N}$ of $\mathscr{C}$. Then we may define the *group completion* of G with regards to the sequence $\{H_r\}_r$ as the quotient group $\mathscr{C}/\mathscr{N}$.

One may check that the quotient groups $\{G/H_r\}_r$ form a directed inverse system and that its direct limit is the group completion of G, e.g. that: $\mathscr{C}/\mathscr{N} = \varprojlim\{G/H_r\}_r$.

If we take $G = \mathbb{Z}$ and H_r to be $p^r\mathbb{Z}$, then we can apply the preceding result since the hypothesis are verified and the group completion of $\mathbb{Z}$ regards to the sequence $p^r\mathbb{Z}$ will be the inverse limit $\varprojlim\{\mathbb{Z}/p^r\mathbb{Z}\}_r$.

Our only task is to check that the notion of group completion and completion for the p-adic metric coincide in that context.

We consider $\{x_n\}_n$, a Cauchy sequence (in the sense of Group completion), then for any $r \geq 0$, we can find N such that $(m,n) \geq N \Rightarrow x_n - xm \in p^r\mathbb{Z}$ and this also imply that $v_p(x_n - x_m) \leq p^{-r}$. Since the p-adic metric is defined by $d_p(x,y) = v_p(x-y)$, this also imply that $d_p(x_n, x_m) \leq p^{-r}$ so that $\{x_n\}_n$ is a fundamental [116] sequence for d_p.

Conversely, let $\{x_n\}_n$ be a fundamental sequence in the p-adic metric space, then for $\varepsilon = p^{-r}$, we can find N such that $(m,n) \geq N \Rightarrow d_p(x_n, x_m) \leq \varepsilon$. This implies that $v_p(x_n, x_m) \leq p^{-r}$ or equivalently that $x_n - x_m \in p^r\mathbb{Z}$ for $(m,n) \geq N$ and this last statement is equivalent to the fact that $\{x_n\}_n$ is a Cauchy sequence for group completion. We have the same result for the null sequences in group completion and in metric spaces.

This prove that the set of Cauchy and null sequence for group completion are isomorphis as groups to, respectively, the set of Fundamental and null sequences in the p-adic metric space.

The completion of a metric space is equivalent to the quotient space of the fundamental sequences by the null sequences so that $\mathbb{Z}_p = \mathscr{C}/\mathscr{N} = \varprojlim\{\mathbb{Z}/p^r\mathbb{Z}\}_r$.

---

[116]We will name, in the context of this exercise, fundamental sequences the Cauchy sequences for metric spaces

64/828 $\mathscr{B}$

Subject: Functor between the category of real vector spaces and the category of complex vector spaces

Every complex vector space can be considered as a real space and every complex linear mapping can be considered as a real linear mapping. Show this correspondence is a covariant functor from the category $L(\mathbb{C})$ of complex vector spaces into the category $L(\mathbb{R})$ of real vector spaces.

SOLUTION:

As indicated in the exercise, we have then a natural correspondence F between $L(\mathbb{C})$ and $L(\mathbb{R})$, defined as follows.

If $E \in L(\mathbb{C})$ and $\{z_\alpha\}_\alpha$ is a basis of E, then we have $z_\alpha = x_\alpha + \mathrm{Im}y_\alpha$ for a family $\{x_\alpha\}$ and a family $\{y_\alpha\}_\alpha$ in $\mathbb{R}$.

If we define from $\{z_\alpha\}_\alpha$ the vectors in $L(\mathbb{R})$: $\{x_\alpha\}$ and $\{x_\alpha^i\}$ with the convention:

$$\mathrm{Im}x_\alpha = x_\alpha^i, \mathrm{Im}x_\alpha^i = -x_\alpha.$$

Then: $F(E) = span\,(x_\alpha) \oplus span\,(x_\alpha^i)$ [117].

That is to say $F(E)$ is an $\mathbb{R}$-vector space and $F \in L(\mathbb{R})$.

The difficult point is to check the functoriality of F. In fact the way F will transform the morphisms is not so obvious.

---

[117] We note $span\,(x_\alpha)$ the vector space spanneded by the family (x_α)

Let $E_1, E_2 \in L(\mathbb{C})$ and $f \in Mor(E_1, E_2) = \mathcal{L}(E_1, E_2)$, if $\{z_{\alpha,1}\}_\alpha$ is a basis of E_1 and if $\{z_{\beta,2}\}_\beta$ is a basis of E_2, then we can define $F : z_{\alpha,1} \to x_{\alpha,1} \oplus x^i{}_{\alpha,1}$ and F: $z_{\beta,2} \to x_{\beta,2} \oplus x^i{}_{\beta,2}$.

Let us suppose that, for $\mathbf{z_1} \in E^1$, $\mathbf{z_1} = \sum_\alpha \lambda_{\alpha,1} z_{\alpha,1}$, then if $\lambda_{\alpha,1} = a_{\alpha,1} + \mathrm{Im} b_{\alpha,1}$, we have:$\mathbf{z_1} = \sum_\alpha (a_{\alpha,1} + \mathrm{Im} b_{\alpha,1}) z_{\alpha,1}$ and $F(\mathbf{z_1}) = \sum_\alpha (a_{\alpha,1} + \mathrm{Im} b_{\alpha,1})(x_{\alpha,1} \oplus x^i{}_{\alpha,1})$ or

$$F(\mathbf{z_1}) = \sum_\alpha (a_{\alpha,1} - b_{\alpha,1}) x_{\alpha,1} \oplus (a_{\alpha,1} + b_{\alpha,1}) x^i{}_{\alpha,1}.$$

So that the action of F is defined by:

$$\mathbf{z_1} = \begin{pmatrix} \cdots \\ a_{\alpha,1} + \mathrm{Im} b_{\alpha,1} \\ \cdots \end{pmatrix} \Rightarrow \begin{pmatrix} \cdots \\ a_{\alpha,1} - b_{\alpha,1} \\ \cdots \end{pmatrix} \oplus \begin{pmatrix} \cdots \\ a_{\alpha,1} + b_{\alpha,1} \\ \cdots \end{pmatrix}$$

Now if $f(z_{\alpha,1}) = \sum_\beta \mu_{\alpha\beta} z_{\beta,2}$ in E_2 with $\mu_{\alpha\beta} = A_{\alpha\beta} + \mathrm{Im} B_{\alpha\beta}$ then:

$$f(\mathbf{z_1}) = \sum_\alpha \lambda_{\alpha,1} f(z_{\alpha,1}) = \sum_\alpha \sum_\beta (a_{\alpha,1} + \mathrm{Im} b_{\alpha,1})(A_{\alpha\beta} + \mathrm{Im} B_{\alpha\beta}) z_{\beta,2}.$$

That is to say:

$$\begin{aligned} f(\mathbf{z_1}) &= \sum_\beta \sum_\alpha (a_{\alpha,1} + \mathrm{Im} b_{\alpha,1})(A_{\alpha\beta} + \mathrm{Im} B_{\alpha\beta}) z_{\beta,2}, \\ &= \sum_\beta \sum_\alpha ((a_{\alpha,1} A_{\alpha\beta} - b_{\alpha,1} B_{\alpha\beta}) + \mathrm{Im}(b_{\alpha,1} A_{\alpha\beta} + a_{\alpha,1} B_{\alpha\beta})) z_{\beta,2}. \end{aligned}$$

This implies that:

$$\begin{aligned} Ff(\mathbf{z_1}) &= \sum_\beta \sum_\alpha (a_{\alpha,1} A_{\alpha\beta} - b_{\alpha,1} B_{\alpha\beta} - b_{\alpha,1} A_{\alpha\beta} - a_{\alpha,1} B_{\alpha\beta}) x_{\beta,2} \\ &\oplus \sum_\beta \sum_\alpha (a_{\alpha,1} A_{\alpha\beta} - b_{\alpha,1} B_{\alpha\beta} + b_{\alpha,1} A_{\alpha\beta} + a_{\alpha,1} B_{\alpha\beta}) x^i{}_{\beta,2} \\ &= \sum_\beta \sum_\alpha ((a_{\alpha,1} - b_{\alpha,1}) A_{\alpha\beta} - (a_{\alpha,1} + b_{\alpha,1}) B_{\alpha\beta}) x_{\beta,2} \\ &\oplus \sum_\beta \sum_\alpha ((a_{\alpha,1} + b_{\alpha,1}) A_{\alpha\beta} + (a_{\alpha,1} - b_{\alpha,1}) B_{\alpha\beta}) x^i{}_{\beta,2}. \end{aligned}$$

To make the calculus simpler, we define new coefficients $X_{\alpha\beta}$ and $Y_{\alpha\beta}$ to be such that:

$$Ff(z_{\alpha,1}) = \sum_{\beta} X_{\alpha\beta} x_{\beta,2} \oplus Y_{\alpha\beta} x^i{}_{\beta,2}.$$

That is to say, we make the following substitution:
$A_{\alpha\beta} - B_{\alpha\beta} = X_{\alpha\beta}$ and:
$A_{\alpha\beta} + B_{\alpha\beta} = Y_{\alpha\beta}.$
Then we have the simplified formula:

$$
\begin{aligned}
Ff(\mathbf{z_1}) \quad =& \quad \sum_{\beta}\sum_{\alpha}(a_{\alpha,1}X_{\alpha\beta} - b_{\alpha,1}Y_{\alpha\beta})x_{\beta,2} \\
\oplus& \quad \sum_{\beta}\sum_{\alpha}(a_{\alpha,1}Y_{\alpha\beta} + b_{\alpha,1}X_{\alpha\beta})x^i{}_{\beta,2}.
\end{aligned}
$$

Then if we put $U_{\alpha,1} = a_{\alpha,1} - b_{\alpha,1}$ and $V_{\alpha,1} = a_{\alpha,1} + b_{\alpha,1}$, we see that Ff transform $\sum_{\alpha} U_{\alpha,1}x_{\alpha,1} \oplus \sum_{\alpha} V_{\alpha,1}x^i{}_{\alpha,1}$ in:

$$
\begin{aligned}
&\sum_{\beta}\sum_{\alpha}((\frac{U_{\alpha,1}+V_{\alpha,1}}{2})X_{\alpha\beta} - (\frac{V_{\alpha,1}-U_{\alpha,1}}{2})Y_{\alpha\beta})x_{\beta,2} \\
\oplus &\sum_{\beta}\sum_{\alpha}((\frac{U_{\alpha,1}+V_{\alpha,1}}{2})Y_{\alpha\beta} + (\frac{V_{\alpha,1}-U_{\alpha,1}}{2})X_{\alpha\beta})x^i{}_{\beta,2}.
\end{aligned}
$$

So that finally, $Ff \in \mathscr{L}(F(E_1), F(E_2))$. From this we still need to check that: $F(g \circ f) = Fg \circ Ff$ for $f \in \mathscr{L}(E_1, E_2)$ and $g \in \mathscr{L}(E_2, E_3)$. We define all the same a basis $z_{\gamma,3}$ for E_3 and the action of F on this basis is to split it in: $z_{\gamma,3} \to x_{\gamma,3} \oplus x^i{}_{\gamma,3}$.

We define $X'_{\beta\gamma}$, $Y'_{\beta\gamma}$, $X_{\alpha\gamma}"$ and $Y_{\alpha\gamma}"$ by:
$Fg(z_{\beta,2}) = \sum_{\gamma} X'_{\beta\gamma} x_{\gamma,3} \oplus Y'_{\beta\gamma} x^i{}_{\gamma,3}$ and:
$F(g \circ f)(z_{\alpha,1}) = \sum_{\gamma} X_{\alpha\gamma}" x_{\gamma,3} \oplus Y_{\alpha\gamma}" x^i{}_{\gamma,3}.$
In order to calculate $Fg \circ Ff$ we make:

$$U_{\beta,2} = \sum_{\alpha} a_{\alpha,1} X_{\alpha\beta} - b_{\alpha,1} Y_{\alpha\beta},$$

$$V_{\beta,2} = \sum_{\alpha} a_{\alpha,1} Y_{\alpha\beta} + b_{\alpha,1} X_{\alpha\beta}.$$

Then: $\frac{U_{\beta,2}+V_{\beta,2}}{2} = \sum_{\alpha} \frac{V_{\alpha,1}}{2} X_{\alpha\beta} + \frac{U_{\alpha,1}}{2} Y_{\alpha\beta}$ and $\frac{U_{\beta,2}-V_{\beta,2}}{2} = \sum_{\alpha} \frac{V_{\alpha,1}}{2} X_{\alpha\beta} - \frac{U_{\alpha,1}}{2} Y_{\alpha\beta}.$

That is to say $(Fg \circ Ff)(\mathbf{z_1})$ is given by the following formula (A):

$$(Fg \circ Ff)(\mathbf{z_1}) =$$

$$\sum_{\gamma}(\sum_{\beta}\sum_{\alpha}(\frac{V_{\alpha,1}X_{\alpha\beta}+U_{\alpha,1}Y_{\alpha\beta}}{2})X'_{\beta\gamma} - (\frac{U_{\alpha,1}X_{\alpha\beta}-V_{\alpha,1}Y_{\alpha\beta}}{2})Y'_{\beta\gamma})x_{\gamma,3}$$

$$\oplus \sum_{\gamma}(\sum_{\beta}\sum_{\alpha}(\frac{V_{\alpha,1}X_{\alpha\beta}+U_{\alpha,1}Y_{\alpha\beta}}{2})Y'_{\beta\gamma} + (\frac{U_{\alpha,1}X_{\alpha\beta}-V_{\alpha,1}Y_{\alpha\beta}}{2})X'_{\beta\gamma})x^i_{\gamma,3}.$$

On the other hand, we know that $f(z_{\alpha,1}) = \sum_{\beta} \mu_{\alpha\beta} z_{\beta,2}$ and $g(z_{\beta,2}) = \sum_{\gamma}\mu'_{\beta\gamma}z_{\gamma,3}$, then:

$(g \circ f)(z_{\alpha,1}) = \sum_{\gamma}(\sum_{\beta})\mu_{\alpha\beta}\mu'_{\beta\gamma})z_{\gamma,3}$, so that if $(g \circ f)(z_{\alpha,1}) = \sum_{\gamma}\mu_{\alpha\gamma}"z_{\gamma,3}$, we have obviously $\mu_{\alpha\gamma}" = \sum_{\beta} \mu_{\alpha\beta}\mu'_{\beta\gamma}.$

Then let us put first $\mu_{\alpha\gamma} = A_{\alpha\gamma} + \mathrm{Im}B_{\alpha\gamma}, \mu'_{\beta\gamma} = A'_{\beta\gamma} + \mathrm{Im}B'_{\beta\gamma}, \mu_{\beta\gamma}" = A_{\alpha\gamma}" + \mathrm{Im}B_{\alpha\gamma}"$. Then we have the following formulas:

$A_{\alpha\gamma}" = \sum_{\beta,2}A_{\alpha\gamma}A'_{\beta\gamma} - B_{\alpha\gamma}B'_{\beta\gamma}$ and:

$B_{\alpha\gamma}" = \sum_{\beta,2}A_{\alpha\gamma}B'_{\beta\gamma} + B_{\alpha\gamma}A'_{\beta\gamma}.$

So that we will have the following formula (B) for $F(g \circ f)(\mathbf{z_1})$:

$$F(g \circ f)(\mathbf{z_1}) =$$

$$\sum_{\gamma}(\sum_{\alpha}(\frac{U_{\alpha,1}+V_{\alpha,1}}{2})(\sum_{\beta}A_{\alpha\beta}A'_{\beta\gamma}-B_{\alpha\beta}B'_{\beta\gamma})$$

$$-\sum_{\alpha}(\frac{V_{\alpha,1}-U_{\alpha,1}}{2})(\sum_{\beta}A_{\alpha\beta}B'_{\beta\gamma}+B_{\alpha\beta}A'_{\beta\gamma})x_{\beta,2}$$

$$\oplus \quad \sum_{\gamma}(\sum_{\alpha}(\frac{U_{\alpha,1}+V_{\alpha,1}}{2})(\sum_{\beta}A_{\alpha\beta}B'_{\beta\gamma}+B_{\alpha\beta}A'_{\beta\gamma})$$

$$+\sum_{\alpha}(\frac{V_{\alpha,1}-U_{\alpha,1}}{2})(\sum_{\beta}A_{\alpha\beta}A'_{\beta\gamma}-B_{\alpha\beta}B'_{\beta\gamma})x^i{}_{\beta,2}.$$

If we make:

$$A_{\alpha\beta} - B_{\alpha\beta} = X_{\alpha\beta},$$
$$A_{\alpha\beta} + B_{\alpha\beta} = Y_{\alpha\beta},$$
$$A'_{\beta\gamma} - B'_{\beta\gamma} = X'_{\beta\gamma},$$
$$A'_{\beta\gamma} + B'_{\beta\gamma} = Y'_{\beta\gamma},$$
$$A_{\alpha\gamma}" - B_{\alpha\gamma}" = X_{\alpha\gamma}",$$
$$A_{\alpha\gamma}" + B_{\alpha\gamma}" = Y_{\alpha\gamma}".$$

Then the formula (B) is turned in (A).

So that $F(g \circ f) = Fg \circ Ff$ and then F is a functor from $L(\mathbb{C})$ to $L(\mathbb{R})$.

65/828

Subject: Functor from $L(\mathbb{R})$ into $L(\mathbb{C})$

Show that the application $L \to \otimes_{\mathbb{R}} \mathbb{C}$ ($\otimes_{\mathbb{R}}$ being the tensor product over $\mathbb{R}$) span a covariant functor from $L(\mathbb{R})$ into $L(\mathbb{C})$.

SOLUTION:

Following the hint, we define the correspondence $F : L(\mathbb{R}) \to L(\mathbb{C})$ by: $F(E) = E \otimes_{\mathbb{R}} \mathbb{C}$, that is to say by the $\mathbb{R}$-tensor product between E and $\mathbb{C}$, $\mathbb{C}$ being seen as a 2-dimensional vector space over $\mathbb{R}$).

The scalar multiplication over $F(E)$ is defined by $z.(a \otimes w) = a \otimes (zw)$ where $(z, w) \in \mathbb{C}$ and $a \in \mathbb{R}$.

Let us check that $E' = F(E)$ is a $\mathbb{C}$-vector space.

By definition, $L \otimes_{\mathbb{R}} \mathbb{C}$ is a $\mathbb{R}$-vector space (see Exercise 61): For u and $v \in L \otimes \mathbb{C}$, we will have $u + v \in L \otimes \mathbb{C}$ so we need only to check the validity of scalar multiplication.

If $u \in L \otimes_{\mathbb{R}} \mathbb{C}$, then $u \in L_1 \boxtimes L_2 / L_1 \odot L_2$ in the sense of Exercise 61. In that context, $L_1 = L$ and $L_2 = \mathbb{C}$ so that L_2 may be viewed as an algebra. If x is a representant of u, then $x = \sum_{i=1}^{n} \lambda_i (a_i \boxtimes b_i)$ for some λ_i's, a_i's and b_i's. We want to define wx when $w \in \mathbb{C}$. If we make $wx = \sum_{i=1}^{n} \lambda_i (a_i \boxtimes wb_i)$ then $wx \in L \boxtimes \mathbb{C}$. Next we need to make sure that wx is independent of the choice of the representant x: Let x' be an other representant of u, then either: 1) $x - x' = (\lambda_1 a_1 + \lambda_2 a_2) \boxtimes b - \lambda_1 (a_1 \boxtimes b) - \lambda_2 (a_2 \boxtimes b)$ or 2): $x - x' = a \boxtimes (\mu_1 b_1 + \mu_2 b_2) - \mu_1 (a \boxtimes b_1) - \mu_2 (a \boxtimes b_2)$.

1) In the first case, $w(x - x') = (\lambda_1 a_1 + \lambda_2 a_2) \boxtimes (wb) - \lambda_1 (a_1 \boxtimes (wb)) - \lambda_2 (a_2 \boxtimes (wb))$.

2) In the second case, $w(x - x') = a \boxtimes (\mu_1(wb_1) + \mu_2(wb_2)) - \mu_1(a \boxtimes (wb_1)) - \mu_2(a \boxtimes (wb_2))$. And in both cases: $w(x - x') \in L \circ \mathbb{C}$, that is to say wx and wx' represent the same element in $L \otimes \mathbb{C}$.

The general case is a linear combination of 1) and 2) so that if $u = \sum_{i=1}^{n} \lambda_i a_i \otimes b_i$ is an element of $L \otimes \mathbb{C}$, then we may write $wu = \sum_{i=1}^{n} \lambda_i a_i \otimes (wb_i) \in L \otimes \mathbb{C}$ for any $w \subset \mathbb{C}$ and this is enough to prove that $L \otimes \mathbb{C}$ is a $\mathbb{C}$ vector space.

Next we must check that F has the functorial properties, e.g. that we have the following situation:

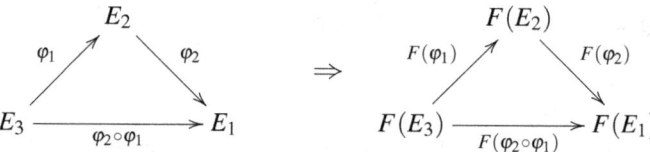

We also need to define precisely what is the action of F on the morphisms of $L(\mathbb{R})$.

But we can build the following diagram, identifying $\mathbb{C}$ and $\mathbb{R}^2$, and we can use the fact that the tensor product $E_1 \otimes_{\mathbb{R}} \mathbb{C}$ is a universal object in the category of bilinear mappings from $E_1 \times \mathbb{C}$ to any vector space F (see Exercise 61):

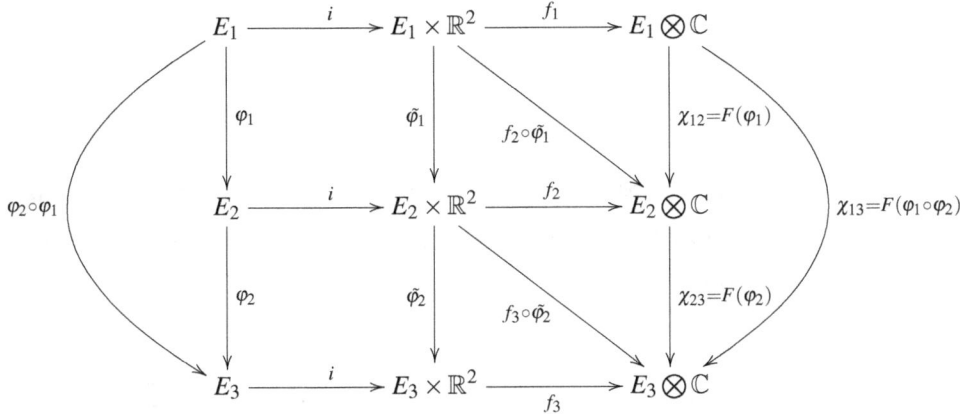

We define i as the canonical injection $E \to E \times \mathbb{R}^2$ and let $\tilde{\varphi}_k$ be the image by i of $\varphi_k, k = 1, 2$. Let f_k be the application uniquely defined betwen $E_k \times \mathbb{C}$ and $E_k \otimes \mathbb{C}, k = 1, 2, 3$ as in Exercise 61. We can define χ_{12}, χ_{23} and χ_{13} such as the above diagram is commutative by using the property of the Tensor Product as a universal object.

If we define $\chi_{12} = F(\varphi_1), \chi_{23} = F(\varphi_2)$ and $\chi_{13} = F(\varphi_2 \circ \varphi_1)$ then $F(\varphi_2 \circ \varphi_1)$ is the unique homomorphism such that $F(\varphi_2 \circ \varphi_1) \circ f_1 = f_3 \circ (\tilde{\varphi}_2 \circ \tilde{\varphi}_1)$. It is immediate to check that $F(\varphi_2) \circ F(\varphi_1)$ fills also that condition so that we must have : $F(\varphi_2 \circ \varphi_1) = F(\varphi_2) \circ F(\varphi_1)$.

66/828

Subject: Non-equivalence between the category of $\mathbb{R}$-vector spaces and the category of $\mathbb{C}$-vector spaces

Show that the categories $L(\mathbb{R})$ and $L(\mathbb{C})$ are not equivalent categories.

SOLUTION:

Let us suppose that categories $L(\mathbb{R})$ and $L(\mathbb{C})$ are equivalent, then we can find a functor F from $L(\mathbb{R})$ to $L(\mathbb{C})$ and a functor G from $L(\mathbb{C})$ to $L(\mathbb{R})$ such that FG=1 in Cov($L(\mathbb{R}),L(\mathbb{R})$) [118] and GF=1 in Cov($L(\mathbb{C}), L(\mathbb{C})$). Then F will define an isomorphism from Aut(A) into Aut(F(A)) for any $\mathbb{R}$-vector space A. Indeed, we see from the following diagram:

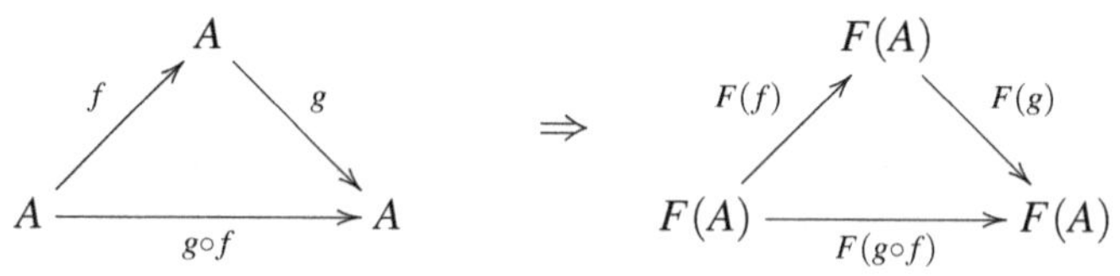

that $F(g \circ f) = F(g) \circ F(f)$.

We can then chose $A = \mathbb{R}$ and so $Aut(A) = GL_1(\mathbb{R}) = (\mathbb{R}^*, \times)$, the real multiplicative group. Let us put that B=F(A) in $L(\mathbb{C})$, then we can consider

---

[118] see [?] or [?] for notations

(whatever B has finite or infinite dimension) the automorphism of B defined by $\varphi_i(\mathbf{z}) = i\mathbf{z}$, we can then see that $\varphi_i^2 = -1_B$. Then if $G = F^{-1}$ we have $G(\varphi_i^2) = G(-1_B) = -G(1_B)$.

It is straightforward to see that $G(1_B) = 1$ (indeed $G(1_B) = G(1_B^2) = G(1_B)^2$ but as $G(1_B) \neq 0$ and $G(1_B) \in \mathbb{R}$ we have $G(1_B) = 1$), so that if $x = G(\varphi_i)$ we must have $x^2 = -1$, which is impossible because $x \in \mathbb{R}$.

67/828

Subject: Equivalent category to the finite dimensional $\mathbb{K}$-vector spaces category

Show that the category of all the finite dimensional $\mathbb{K}$-vector spaces is equivalent to one of its subcategory containing an enumerable set of objects.

SOLUTION:

We define a correspondence F between the category $\mathcal{C}VF$ of finite-dimensional $\mathbb{K}$-vector spaces and its subcategory $\mathcal{C}VF'$ made with the spaces $\{\mathbb{K}^n\}_{n \in \mathbb{N}}$ as follows: we chose for each objects E in $\mathcal{C}VF$ a basis $\{e_1^E, \ldots, e_n^E\}$ and we define a bijective linear mapping $f_E : E \to \mathbb{K}^n$ by $f(e_i^E) = \{\delta_{ij}\}_{j=1}^{j=n} \in \mathbb{K}^n$, with the convention that $\delta_{ij} = 0$ if $i \neq j$ and $\delta_{ii} = 1$. We define the action of a functor F on the objects of $\mathcal{C}VF$ by: $F(E) = f_E(E)$, that is to say $F(E) = \mathbb{K}^{dim(E)}$.

Morphisms in the category $\mathcal{C}VF$ are linear mappings between vector spaces (e.g. $Mor(E,F) = \mathscr{L}(E,F)$). We define the action of F on these morphisms by : $F(\varphi)f_E(x) = f_F \circ \varphi(x)$ for any $\varphi \in Mor(E,F), x \in E$.

F has functorial properties, indeed if $f \in \mathscr{L}(E,F)$ and $g \in \mathscr{L}(F,G)$ where E, F and G are objects in $\mathcal{C}VF$ such that $dim(E) = n$, $dim(F) = m$, $dim(G) = q$. If $A = A_{nm}$ and $B = B_{mq}$ are the matrixes of respectively f and g when E,F and G are provided respectively with basis e^E, e^F and e^G then the matrix of $g \circ f$ will be $AB = (AB)_{nq}$. As the matrix of $F(f)$ and $F(g)$ will be also respectively A and B, then the matrix of $F(g) \circ F(f)$ will be AB that is to say the matrix of $F(g \circ f)$ so that $F(g \circ f) = F(g) \circ F(f)$.

We can define F^{-1} by the reverse process without ambiguity, because all finite dimensional vector spaces over the same field $\mathbb{K}$ with the same dimension are isomorphic so that the choice of the basis e^E does not matter.

68/828

Subject: Equivalent category to the finite group category

Show that the category of finite group is equivalent to one of its subcategory containing an enumerable set of objects.

SOLUTION:

This exercise is similar to the Exercise 67 except that we consider finite groups instead of vector spaces.

We note S_n the permutation group of n elements $S_n = \{\sigma_i^n\}_{i=1}^{i=n!}$. The set of objects $\{S_n\}$ leads to a subcategory $\mathfrak{CGG}$ of the category $\mathfrak{CG}$ of finite groups and $card(\mathfrak{CGG}) = card(\mathbb{N})$. Our goal is to find a functor $F : \mathfrak{CG} \to \mathfrak{CGG}$.

First try.

We cannot associate directly a group G_n ($\#G_n = n$) with the permutation groups S_n because we could not describe precisely how F would transform $Mor(G_n, G_p)$ for $(n, p) \in \mathbb{N}$. Instead we can use the fact that S_n contains cyclic subgroups, namely the group G_{σ_n} generated by the cyclic permutations σ_n defined by $\sigma_n(1,2,3,\ldots,n) = (2,3,\ldots,n,1)$, σ_n is of order n: $\sigma_n^n = 1$. So that, if we have $G_n = \{1, x_1, \ldots, x_{n-1}\}$, we may define the action of F on the objects of $\mathfrak{CG}$ by: $F(1) = 1$, $F(x_i) = \sigma_n^i$. Then we have a correspondence $G_n \to G_{\sigma_n}$.

Now if φ is an homomorphism $G_n \to G_p$, then $\#\ker\varphi\,|\,\#G_n = n$ and $\#\text{Im } \varphi = \frac{\#G_n}{\#\ker\varphi} = m$, and $m|p$ [119].

So that we may define $F\varphi : G_{\sigma_n} \to G_{\sigma_p}$ by: $F\varphi(\sigma) = \sigma^{\#\ker\varphi} = \sigma^{\frac{n}{m}}$ and then $\#F\varphi(G_{\sigma_n}) = m \times \frac{n}{m} = n$.

We must next, prove that F has the functorial properties, e.g. that the following diagram is commutative:

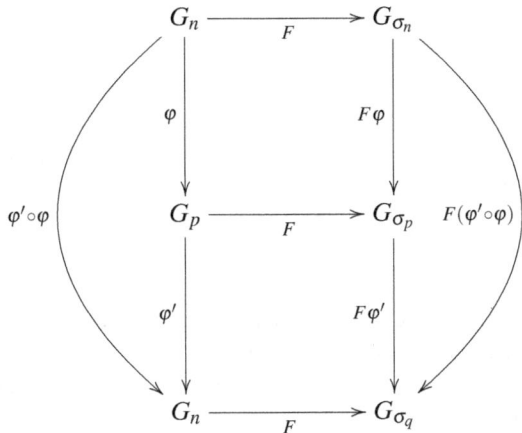

$F\varphi$ is defined by: $\sigma \to \sigma^{n/m}, m = \#\text{Im } \varphi, n/m = \#\ker\varphi, m|p$.
$F\varphi'$ is defined by: $\sigma \to \sigma^{p/r}, m = \#\text{Im } \varphi', p/r = \#\ker\varphi', r|q$.
Thus $F\varphi' \circ F\varphi$ is defined by: $\sigma \to \sigma^{(n/m)(p/r)}$.
On the other hand, $F(\varphi \circ \varphi')$ is defined by:
$\sigma \to \sigma^{n/s}, s = \#\text{Im } (\varphi' \circ \varphi), n/s = \#\ker(\varphi' \circ \varphi), s|q$.

But $\#\text{Im } (\varphi' \circ \varphi) = \#\text{Im } (\varphi') \times \frac{\#\ker\varphi'}{p} = r \times \frac{m}{p}$ so that $s = \frac{r \times m}{p}$ and $\frac{s}{q} = \frac{m}{p} \times \frac{r}{q} \in \mathbb{N}$.

This show that $F(\varphi' \circ \varphi) = F\varphi' \circ F\varphi$. Then F is a functor from $\mathfrak{CG}$ to $\mathfrak{CGG}$ and in fact, a functor from $\mathfrak{CG}$ to $\mathfrak{CCG}$ if we note $\mathfrak{CCG}$ to be the category of cyclic groups of finite orders.

---

[119] Since $\text{Im } \varphi \sim G_n/\ker\varphi$ and $\text{Im } \varphi$ is a subgroup of G_p.

The problem with our construction is that we associate all groups of order n with a single group so that F is a Functor but we cannot define F^{-1}. Indeed, from $\{1, \sigma, \ldots, \sigma^{n-1}\}$ we can associate conversely many different groups G_n:. Let $\mathscr{G}_n$ be the set of finite groups of order n and let $s_n = \#\mathscr{G}_n$. If $n > 3$. For $n \geq 3$, $s_n > 1$. For example we consider the case $n = 4$. We define G_4^1 and G_4^2 two different finite groups in $\mathscr{G}_4$ by the following tables $M(G_4^1)$ and $M(G_4^2)$:

$$G_4^1 : \begin{pmatrix} 1 & x_1 & x_2 & x_3 \\ x_1 & 1 & x_3 & x_2 \\ x_2 & x_3 & 1 & x_1 \\ x_3 & x_2 & x_1 & 1 \end{pmatrix} \quad \text{and:} \quad G_4^2 : \begin{pmatrix} 1 & x_1 & x_2 & x_3 \\ x_1 & x_2 & x_3 & 1 \\ x_2 & x_3 & 1 & x_1 \\ x_3 & 1 & x_1 & x_2 \end{pmatrix}$$

(where the matrix are being defined by $x_{ij} = x_i x_j$, $x_0 = 1$).

These two groups are not isomorphic to each other and then $s_4 \geq 2$. So that our construction in not valid.

Solution.

We note that the table of multiplication (or table of law of composition) $M(G_n)$ of a group G_n will define it uniquely.

Furthermore the rows (or columns) of this $n \times n$ matrix are permutations of the set $\{0, 1, \ldots, n-1\}$.

$$\text{Indeed } M(G_n) = \begin{pmatrix} & \vdots & \\ \cdots & x_i x_j & \cdots \\ & \vdots & \end{pmatrix}$$

and $x_i x_j = x_{k_{ij}}$ for some $k_{ij} \in \{0, 1, \ldots, n-1\}$ or equivalently $x_i x_j = x_{\varphi_i(j)}$. φ_i will be a mapping from $\{0, 1, \ldots, n-1\}$ to $\{0, 1, \ldots, n-1\}$ and this will be an injective mapping since $\varphi_i(j) = \varphi_i(j') \Leftrightarrow x_j = x_{j'} \Leftrightarrow j = j'$. So that φ_i is a permutation $\sigma_i^n \in S_n$ for any $i \in \mathbb{N}, i \leq n-1$.

Besides the set of rows (or columns) $\{\sigma_i^n\}_{i=1}^{i=n}$ will form a subgroup of S_n. Indeed, if $(p, q) \in \{0, 1, \ldots, n-1\}$ then $x_k x_q = x_{\sigma_q(k)}, \forall k \in \{0, 1, \ldots, n-1\}$. So that $x_{\sigma_q(k)} x_p = x_{\sigma_p \circ \sigma_q(k)}$ $(= x_{\sigma_p(\sigma_q(k))})$ but we have also $x_{\sigma_q(k)} = x_k x_q$ so that $x_k x_q x_p = x_{\sigma_p \circ \sigma_q(k)}$ or $x_k(x_q x_p) = x_{\sigma_p \circ \sigma_q(k)}$ or:

$$x_{\sigma_{p(q)}(k)} = x_{\sigma_p \circ \sigma_q(k)}, \forall k \in \{0, 1, \ldots, n-1\}.$$

That is to say:

$$x_{\sigma_{\sigma_p(q)}} = x_{\sigma_p \circ \sigma_q}$$

and we must have the additional conditions[120]: $\sigma_0 = 1_n; \sigma_i(j) = 0, \forall j$
then $\forall p, \exists q / g_p \circ \sigma_q = 1_n (q = \sigma_p^{-1}(0))$.

This shows that the σ_i's are a subgroup of order n of S_n.

Examples:

The groups G_4^1 and G_4^2 which we have defined previously have multiplication tables whose rows are permutations[121]:

$$M(G_4^1) = \begin{pmatrix} 1_4 \\ (0,1)(2,3) \\ (0,3)(1,2)(0,1)(2,3) \\ (0,1)(2,3)(0,3)(1,2)(0,1)(2,3) \end{pmatrix}$$

So that $F(G_4^1)$ is made up with the following elements:

$$\{1_4; (0,1)(2,3); (0,3)(1,2)(0,1)(2,3); (0,1)(2,3)(0,3)(1,2)(0,1)(2,3)\}$$

and thus it is a subgroup of S_4.

$$M(G_4^2) = \begin{pmatrix} \sigma^0 \\ \sigma^1 \\ \sigma^2 \\ \sigma^3 \end{pmatrix}$$

where σ is the cyclic permutation $(0,1,2,3) \to (3,0,1,2)$
and $F(G_4^2) = \{1, \sigma, \sigma^2 \sigma^3\} = (\sigma) \in S_4$.

We can try two other groups, like:

1) The additive group $(\mathbb{Z}/6\mathbb{Z}, +)$.

---

[120] 1_n being the identical permutation in S_n
[121] (i,j) is the permutation that permute i and j and left the rest unchanged

$$M((\mathbb{Z}/6\mathbb{Z},+)) = \begin{pmatrix} 0 & 1 & 2 & 3 & 4 & 5 \\ 1 & 2 & 3 & 4 & 5 & 0 \\ 2 & 3 & 4 & 5 & 0 & 1 \\ 3 & 4 & 5 & 0 & 1 & 2 \\ 4 & 5 & 0 & 1 & 2 & 3 \\ 5 & 0 & 1 & 2 & 3 & 4 \end{pmatrix}.$$

So that if σ_6 is the cyclic permutation $(0,1,2,3,4,5) \to (1,2,3,4,5,0)$ then
$$F : (\mathbb{Z}/6\mathbb{Z},+) \to \{1, \sigma_6, \ldots, \sigma_6^5\} = (\sigma) \in S_6.$$

2) The multiplicative group $(\mathbb{Z}/5\mathbb{Z}^*, \times)$:

$$M((\mathbb{Z}/5\mathbb{Z}^*, \times)) = \begin{pmatrix} 1 & 2 & 3 & 4 \\ 2 & 4 & 1 & 3 \\ 3 & 1 & 4 & 2 \\ 4 & 3 & 2 & 1 \end{pmatrix}$$

then $\sigma_0^4 = 1_4$,

$\sigma_1^4 : (0,1,2,3) \to (1,3,0,2)$,

$\sigma_2^4 : (0,1,2,3) \to (2,0,3,1)$,

$\sigma_3^4 : (0,1,2,3) \to (3,2,1,0)$,

and $F : (\mathbb{Z}/6\mathbb{Z},+) \to \{\sigma_0^4, \sigma_1^4, \sigma_2^4, \sigma_3^4\}$.

Now we need to check the functoriality of F: for this we consider three finite groups: G_n, H_m and K_q of orders, respectively, n, m and q. We define the action of F on these groups by:

$G_n \to \{\sigma_i^n\}_{i=0}^{i=n-1}$,

$H_m \to \{\sigma_u^r\}_{u=0}^{u=r-1}$,

$K_q \to \{\sigma_u^q\}_{a=0}^{a=q-1}$.

Let φ an homomorphism $G_n \to H_m$ and ϕ an homomorphism $H_m \to K_q$.

We can identify H_m and $G_n/\ker\varphi$: $Im\varphi \approx G_n/\ker\varphi$ and in the same way: $Im\phi \approx H_m/\ker\phi$, $Im(\phi \circ \varphi) \approx G_n/\ker(\phi \circ \varphi)$. Indeed we can restrict ourselves to the images of these homomorphism and consider that they are surjective so that $K_q \subset H_m \subset G_n$.

On the other hand, $F\varphi$ is defined as the rows of the law of multiplication $M(G_n/\ker\varphi)$ e.g. if $\{\tilde{x}_u\}_{u=0}^{u=r-1} = G_n/\ker\varphi$ then $\sigma_u^r(v) = \tilde{x}_u\tilde{x}_v, u = 0, \ldots, r-1$, $r = \#(G_n/\ker\varphi)$ $r|n$ the set $\{\sigma_u^r\}_{u=0}^{u=r-1}$ is a subgroup of $F(H_m)$

in S_m, e.g. we have a chain of subgroups $\{\sigma_u^r\}_{u=0}^{u=r-1} \subset F(H_m) \subset H_m$.

$F\phi$ can be defined identically by the rows of the law of composition in $H_m/\ker\phi$ and again we can define $F(\phi \circ \varphi)$ by the rows of the law of composition of $G_n/\ker(\phi \circ \varphi)$ so that in order to show the functoriality of the mapping F we need only to show that $H_m/\ker\phi$ and $G_n/\ker(\phi \circ \varphi)$ are isomorphic but this is equivalent to: $(G_n/\ker\varphi)/\ker\phi = G_n/\ker(\phi \circ \varphi)$ and the demonstration of this result is straightforward from the basic results of quotient groups.

The only task left is to check that F^{-1} is well-defined. The image of F is made of the set of the subgroups G_n^S of the symmetric group S_n, for $n \geq 0$ such that:

1) $\#G_n^S = n$.
2) $\forall \sigma \in G_n^S, \sigma(0) = 0$ (e.g. 0 is a fixed point).

For any group $G_n^S = \{\sigma_i^n\}$ in $Im(F)$, we define the group $G_n = F^{-1}(G_n^S)$ by a set of elements[122] $\{1, x_1, \ldots, x_n\}$ with the following law of multiplication:

$$x_i \times x_j = x_{\sigma_i(j)}$$

and from this we can see that:

- $\forall x_i, x_j \in G_n, x_i \times x_j \in G_n$,
- $\forall x_i \exists x_j (j = (\sigma_i^n)^{-1}(0))/x_i \times x_j = 1$,
- $\forall x_i, 1 \times x_i = x_i \times 1 = 1$.

This shows that G_n is a group of order n. As G_n is uniquely defined by its law of composition, F^{-1} is well-defined (F being a functor so is F^{-1}). So that the category of finite groups is equivalent to the subcategory of permutation groups.

---

[122] $x_0 = 1$ and $\sigma_0^n = 1_n$

69/828 $\mathscr{B}$

Subject: Group algebra $\mathbb{K}[G]$

Let G be a group and $\mathbb{K}$, a field. We consider the set $\mathbb{K}[G]$ of formal linear combination of elements with coefficients in $\mathbb{K}$. $\mathbb{K}[G]$ is an algebra for addition, multiplication by elements of $\mathbb{K}$ and for product.

Show that:

a) The correspondence $G \to \mathbb{K}[G]$ is a covariant functor from the group category into the $\mathbb{K}$-Algebras category;

b) The application $G \to \mathbb{K}[G]$ is a universal object from the category of multiplicative applications from the group G into $\mathbb{K}$-Algebra's.

SOLUTION:

We then have $\mathbb{K}[G] = \{\sum_\alpha \lambda_\alpha g_\alpha, \lambda_\alpha \in \mathbb{K}, g_\alpha \in G\}$.

a) We consider the correspondence $F : G \to \mathbb{K}[G]$ defined that way: F is a correspondence from the objects of the category $\mathfrak{CG}$ to the objects of the category of $\mathbb{K}$-algebra's.

If we define the action of F on the morphisms of $\mathfrak{CG}$ by: $\varphi \in Mor(G_1, G_2) \Rightarrow F\varphi(\sum_\alpha \lambda_\alpha g_\alpha) = \sum_\alpha \lambda_\alpha \varphi(g_\alpha)$, then this defines $F\varphi$ as a morphism $\mathbb{K}[G_1] \to \mathbb{K}[G_2]$.

Let us check that F is a functor, e.g. that the following diagram is commutative:

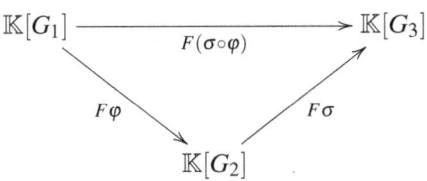

$$
\begin{aligned}
F\varphi\left(\sum_\alpha \lambda_\alpha g_\alpha\right) &= \sum_\alpha \lambda_\alpha \varphi(g_\alpha) \text{ and } F(\sigma \circ \varphi)\left(\sum_\alpha \lambda_\alpha g_\alpha\right), \\
&= \sum_\alpha \lambda_\alpha (\sigma \circ \varphi)(g_\alpha), \\
&= F\sigma\left(\sum_\alpha \lambda_\alpha \varphi(g_\alpha)\right), \\
&= (F\sigma \circ F\varphi)\left(\sum_\alpha \lambda_\alpha g_\alpha\right).
\end{aligned}
$$

So that F has the functorial properties.

b) We consider the category $\mathfrak{C}[G]$ of multiplicatives applications from G to $\mathbb{K}$-algebra's: this category is made with pairs $(\mathbb{A}, \varphi)$ where φ is a multiplicative application $G \to \mathbb{A}$ and $\mathbb{A}$ is an algebra. The morphisms from two objects $(\mathbb{A}, \varphi)$ and $(\mathbb{B}, \phi)$ are the algebra-homomorphisms $\chi : \mathbb{A} \to \mathbb{B}$ such that the following diagram be commutative:

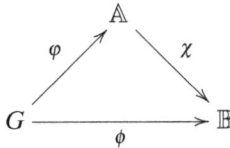

This defines a category, indeed, we have the following diagram:

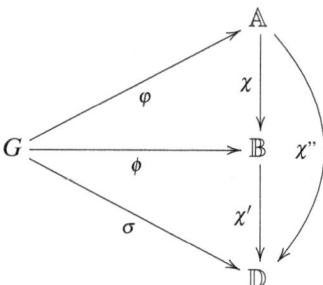

Where $\mathbb{A}$, $\mathbb{B}$ and $\mathbb{D}$ are $\mathbb{K}$-algebra's. χ and χ' are two morphism from, respectively, $Mor((\mathbb{A}, \varphi); (\mathbb{B}, \phi))$ and $Mor((\mathbb{B}, \phi); (\mathbb{D}, \sigma))$ Then $\chi" = \chi' \circ \chi$ will be a morphism from $Mor((\mathbb{A}, \varphi); (\mathbb{D}, \sigma))$, indeed, $\chi' \circ \chi$ is an algebra-homomorphism: $\mathbb{A} \to \mathbb{D}$ and $\chi' \circ \chi \circ \varphi = \chi' \circ (\chi \circ \varphi) = \chi' \circ \phi = \sigma$.

Then $(\mathbb{K}[G], f)$ will be a universal object, if we note f the correspondence $G \to \mathbb{K}[G]$, $g_\alpha \to g_\alpha$. Indeed, let $(\mathbb{A}, \varphi)$ be an object in the category, we define χ by $\chi(g_\alpha) = \varphi(g_\alpha)$, $\forall \alpha$. Then:

$$\chi\left(\sum_\alpha \lambda_\alpha g_\alpha\right) = \sum_\alpha \lambda_\alpha \chi(g_\alpha) = \sum_\alpha \lambda_\alpha \varphi(g_\alpha).$$

From its definition, if χ exist, the following diagram will be commutative:

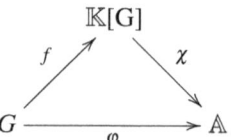

We check that χ is an algebra-homomorphism.
If $x = \sum_\alpha \lambda_\alpha g_\alpha$ and $y = \sum_\beta \mu_\beta g_\beta$ then:

$$
\begin{aligned}
\chi(xy) &= \chi\left(\left(\sum_{\alpha}\lambda_{\alpha}g_{\alpha}\right)\left(\sum_{\alpha}\mu_{\beta}g_{\beta}\right)\right), \\
&= \sum_{\alpha\beta}\lambda_{\alpha}\lambda_{\beta}\chi(g_{\alpha}g_{\beta}), \\
&= \sum_{\alpha\beta}\lambda_{\alpha}\lambda_{\beta}\varphi(g_{\alpha}g_{\beta}), \\
&= \sum_{\alpha\beta}\lambda_{\alpha}\lambda_{\beta}\varphi(g_{\alpha})\varphi(g_{\beta}), \\
&= \sum_{\alpha\beta}\lambda_{\alpha}\lambda_{\beta}\chi(g_{\alpha})\chi(g_{\beta}), \\
&= \left(\sum_{\alpha}\lambda_{\alpha}\chi(g_{\alpha})\right)\left(\sum_{\beta}\lambda_{\beta}\chi(g_{\beta})\right), \\
&= \chi\left(\sum_{\alpha}\lambda_{\alpha}g_{\alpha}\right)\chi\left(\sum_{\beta}\lambda_{\beta}g_{\beta}\right), \\
&= \chi(x)\chi(y).
\end{aligned}
$$

So that χ is unique and well-defined as an Algebra-homomorphism, thus $(\mathbb{K}[G], f)$ is the requested universal object.

List of Figures

Appendix

Perl source code for Exercise 42

A Perl 5 code source for building the bijection.

```perl
use strict;
use Clone qw(clone);
require SPECIAL::B;
require SPECIAL::I;


my    $MAX_LEVEL=5;
my    @DA_2;
my    @DA_3;
our   $VARDBG;
our   $CNT;
our   $PNT;

my $B0_2=B->new(1);
$B0_2->SETP(2);
$B0_2->SETELEMENT(1,I->new(0,1));

my $B0_3=B->new(1);
$B0_3->SETP(3);
$B0_3->SETELEMENT(1,I->new(0,1));
```

273

```perl
########################################
#  PROCESSB CALLBACK
########################################
my $PROCESSB=sub
{


my $B=shift;

  if($B->ISACTIVE()!= 1)
  {
  return;
  }

my $N=$B->GETNELEMENTS;


  if($N>1)
  {
  our $B1=B->new(1);
  our $B2=B->new($N-1);


  my $I1 =$B->GETELEMENT(1);
  $B1->SETELEMENT(1,$I1);

    for($CNT=1;$CNT<$N;$CNT++)
    {
    my $I_C=$B->GETELEMENT($CNT+1);
    $B2->SETELEMENT($CNT,$I_C);
    }

  $B->SETCHILD($B1,0);
```

```
$B->SETCHILD($B2,1);

$B->SETACTIVE(0);

$B1->SETFATHER($B);
$B2->SETFATHER($B);

my $I2=$B->GETELEMENT(2);
my $J=I->new($I1->GETY,$I2->GETX);

$B->SETDELTA($J);
}

elsif($N==1)
{

my $P=$B->GETP();
our $B1=B->new(1);
our $B2=B->new($P-1);

my $I=$B->GETELEMENT(1);

my @II=CSPLIT($I,$P);

$B1->SETELEMENT(1,$II[0]);

   for($CNT=1;$CNT<$P;$CNT++)
   {
   $B2->SETELEMENT($CNT,$II[$CNT]);
   }

$B->SETCHILD($B1,0);
$B->SETCHILD($B2,1);

$B->SETACTIVE(0);
```

```perl
$B1->SETFATHER($B);
$B2->SETFATHER($B);

my $J=I->new($II[0]->GETY,$II[1]->GETX);

$B->SETDELTA($J);

}

else
{
die("Fatal error \$N out of bounds");
}

};


#####################################
#  CSPLIT
#####################################

sub CSPLIT
{

my ($I,$P)=@_;
my $X=$I->GETX;
my $Y=$I->GETY;
my @II_;

my $D=($Y-$X)/(2*$P-1);

  for($CNT=0;$CNT<$P;$CNT++)
  {
  my $INEW = new I($X+(2*$CNT*$D),$X+((2*$CNT+1)*$D));
```

```
 push @II_,clone($INEW);
 undef $INEW;
 }


return @II_;

}


######################################
#   GETBFROMBINARY
######################################

sub GETBFROMBINARY
{
my ($B0,$N)=@_;
my $B=$B0;


my $K= length sprintf("%b",$N);

if($K<2)
{
die("\$K out of bounds");
}

my $BIN;

  for($CNT=$K-2;$CNT>=0;$CNT--)
  {
  $BIN=($N >> $CNT )& 1;
  $B=$B->GETCHILD($BIN);
  }
```

```perl
return $B;

}




#####################################
#   BROWSEBTREE
#####################################

sub BROWSEBTREE
{
my $B0=shift;
my $LEV=shift;
my $F=shift;
my (@ARGS)=@_;

my $N=1;
my $B=$B0;
my $lv;

if($B==undef)
{
die("\$B == NULL");
}

   while($lv<$LEV)
   {

     $F->($B,@ARGS);

$N++;
$B=GETBFROMBINARY($B0,$N);
$lv=int(log($N)/log(2));
```

```
  }

}

#####################################
#   BUILDDELTAARRAY CALLBACK
#####################################

my $BUILDELTAARRAY =sub
{

my $B=shift;
$PNT=@_[0];

push  @{$PNT} , $B->GETDELTA();

};


################# MAIN ##############

  BROWSEBTREE($BO_2,$MAX_LEVEL,$PROCESSB);
  BROWSEBTREE($BO_3,$MAX_LEVEL,$PROCESSB);

  my @ARGZ;
  $ARGZ[0]=\@DA_2;
  BROWSEBTREE($BO_2,$MAX_LEVEL,$BUILDELTAARRAY,@ARGZ);
  $ARGZ[0]=\@DA_3;
  BROWSEBTREE($BO_3,$MAX_LEVEL,$BUILDELTAARRAY,@ARGZ);

  #associate intervals of @DA_2 with intervals of @DA_3 for scilab
  open FILE , ">scilab.sc" or die("CANNOT OPEN FILE scilab.sc FOR WRITING'

  for($CNT=0;$CNT<$#DA_2;$CNT++)
```

```
  {
   print FILE 'plot2d(['.$DA_2[$CNT]->GETX.','.$DA_2[$CNT]->GETY
   ['.$DA_3[$CNT]->GETX.','.$DA_3[$CNT]->GETY.']);'."\n";
  }
  close FILE;


########################################
#  LOG
########################################
sub LOG()
{

  my $inn=shift;
  my $err_flag=shift;
  my $ERROR="";
  if($err_flag==1)
  {
$ERROR="Fatal Error in";
  }

  print $ERROR."DBG=>",(caller(1))[3],"....",$inn,"\n";

}


########################################
#  PRINTARRAYOFI
########################################
sub PRINTARRAYOFI()
{

  my (@arr)=@_;
  my $elt;
  my $buf="number of elements in array of I = ".($#arr+1)."\n";
  foreach $elt(@arr)
```

```perl
    {
    $buf=$buf.$elt->DUMP;
    }

return $buf;
}

1;

package B;
use strict;

#####################################
#     OBJECT B
#####################################


sub new
{


my $INVOCANT=shift;
my $CLASS= ref($INVOCANT)||$INVOCANT;


my $N = shift;

   my $SELF={
   NUMBER_OF_ELEMENTS=> $N,
   ISACTIVE=>1,
   ELEMENTS=>undef,
   DELTA=>0,
   P=>0,
   CHILDS=>undef,
   FATHER=>undef,
```

```perl
    @_
    };

return bless $SELF,$CLASS;
}


sub GETNELEMENTS
{
my $SELF=shift;

return $SELF->{NUMBER_OF_ELEMENTS};
}

sub GETCHILD
{
my $SELF=shift;
my $POS=shift;
return $SELF->{CHILDS}[$POS];
}

sub SETCHILD
{
my $SELF=shift;
my $B=shift;
my $POS=shift;

$SELF->{CHILDS}[$POS]=$B;
}

sub SETFATHER
{
my $SELF=shift;
my $B=shift;
$SELF->{FATHER}=$B;
```

```perl
$SELF->{P}=$B->GETP();
}

sub GETFATHER
{
  my $SELF=shift;
return $SELF->{FATHER};
}

sub SETELEMENT
{
  my $SELF=shift;
my $N=shift;
my $I=shift;

if(($N>$SELF->{NUMBER_OF_ELEMENTS})||($N<1))
{
die("\$N=$N outbounds");
}

$SELF->{ELEMENTS}[$N-1]=$I;
}

sub GETELEMENT
{
  my $SELF=shift;
my $N=shift;

if(($N>$SELF->{NUMBER_OF_ELEMENTS})||($N<1))
{
die("\$N=$N outbounds");
}

return $SELF->{ELEMENTS}[$N-1];
}
```

```perl
sub SETACTIVE
{
  my $SELF=shift;
my $BOOL=shift;

if(($BOOL!=0)&&($BOOL!=1))
{
die("\$BOOL=$BOOL not a boolean value");
}

$SELF->{ISACTIVE}=$BOOL;
}

sub ISACTIVE
{
  my $SELF=shift;
return $SELF->{ISACTIVE};
}

sub GETDELTA
{
  my $SELF=shift;
return $SELF->{DELTA};
}

sub SETDELTA
{
  my $SELF=shift;

my $D=shift;
$SELF->{DELTA}=$D
}
```

```perl
sub SETP
{
  my $SELF=shift;
my $P=shift;
$SELF->{P}=$P;
}

sub GETP
{
  my $SELF=shift;
return $SELF->{P};
}

sub DUMP
{
    my $SELF=shift;

    my $buf="\n DUMP B ------>\n";

    foreach my $key (sort(keys %$SELF))
    {
      $buf=$buf.sprintf ($key.'='.$SELF->{$key}."\n");
    }
    $buf=$buf."<------ DUMP B\n";
return $buf;
}

1;

package I;
use strict;

#####################################
#    OBJECT I
#####################################
```

```perl
sub new
{


my $INVOCANT=shift;
my $X_=shift;
my $Y_=shift;
my $CLASS= ref($INVOCANT)||$INVOCANT;
my $SELF={
X=>$X_,
Y=>$Y_,
@_
};
return bless $SELF,$CLASS;
}


sub GETX()
{
  my $SELF=shift;
  return $SELF->{X};

}


sub GETY()
{
    my $SELF=shift;
  return $SELF->{Y};

}
```

```perl
sub SETX()
{
  my $SELF=shift;
  my $X=shift;
  $SELF->{X}=$X;

}


sub SETY()
{
    my $SELF=shift;
  my $Y=shift;
  $SELF->{Y}=$Y;

}

sub DUMP
{
    my $SELF=shift;
    my $buf="\n DUMP I ------>\n";
    foreach my $key (sort(keys %$SELF)) {
                $buf=$buf.sprintf ($key.'='.$SELF->{$key}."\n");
    }
      $buf=$buf."<------ DUMP I\n";
return $buf;
}

1;
```

Index

Bibliography

[1] N. Bourbaki. *Algebra II*. Springer , U.S.A, 2003.

[2] N. Bourbaki. *Elements of mathematics. Lie groups and Lie algebras ,
Volume 3 to Volume 7*. Springer , U.S.A, 2005.

[3] J. Dixmier. *Algèbres Enveloppantes*. Gauthier-Villars , France, 1974.

[4] A. K. et A.D. Gvishiani. *Théorèmes et Problèmes d'Analyse Fonc-
tionnelle*. Mir-Nauka , éditions de Moscou , U.R.S.S, 1979.

[5] R. Godement. *Introduction a la théorie des Groupes de Lie*. Springer-
Verlag , France, 2003.

[6] N. Jacobson. *Lie Algebras*. Dover , U.S.A, 1979.

[7] A. Karazishvili. *Strange Functions in Real Analysis*. Chapman
Hall/CRC , U.S.A, 2005.

[8] S. Katok. *Real And p-adic analysis courses notes for MATH 497C*.
The Pennsylvania State University , U.S.A, 2000.

[9] A. Kirillov and A. Gvishiani. *Theorems and Problems in Functional
Analysis*. Springer-Verlag , U.S.A, 1982.

[10] S. Lang. *Algebra (Revised third edition)*. Springer-Verlag , U.S.A,
2002.

[11] J.-P. Serre. *Lie Algebras and Lie Groups.* W.A Benjamin Inc , The Netherlands, 1965.